Laboratory Manual

Chemistry in Context

Applying Chemistry to Society

Seventh Edition

Edited by

Jennifer A. Tripp
San Francisco State University

Mc Graw Hill
Connect
Learn
Succeed™

The McGraw·Hill Companies

Connect
Learn
Succeed™

LABORATORY MANUAL TO ACCOMPANY CHEMISTRY IN CONTEXT:
APPLYING CHEMISTRY TO SOCIETY, SEVENTH EDITION

2 3 4 5 6 7 8 9 0 QDB/QDB 1 0 9 8 7 6 5 4 3 2

ISBN 978–0–07–733448–2
MHID 0–07–733448–5

Vice President, Editor-in-Chief: *Marty Lange*
Vice President, EDP: *Kimberly Meriwether David*
Senior Director of Development: *Kristine Tibbetts*
Publisher: *Ryan Blankenship*
Sponsoring Editor: *Todd L. Turner*
Developmental Editor: *Jodi Rhomberg*
Executive Marketing Manager: *Tamara L. Hodge*
Project Coordinator: *Mary Jane Lampe*
Senior Buyer: *Kara Kudronowicz*
Designer: *Tara McDermott*
Cover Designer: *Christopher Reese*
Cover Image: © *Getty Images/Grant Faint*
Compositor: *Aptara, Inc.*
Typeface: *10/14 Times Roman*
Printer: *Quad/Graphics*

Some of the laboratory experiments included in this text may be hazardous if materials are handled improperly or if procedures are conducted incorrectly. Safety precautions are necessary when you are working with chemicals, glass test tubes, hot water baths, sharp instruments, and the like, or for any procedures that generally require caution. Your school may have set regulations regarding safety procedures that your instructor will explain to you. Should you have any problems with materials or procedures, please ask your instructor for help.

Contents

Preface to Instructors

This laboratory manual accompanies the seventh edition of *Chemistry in Context, Applying Chemistry to Society*. Both the text and this laboratory manual are designed for college students majoring in disciplines outside of the natural sciences. *Chemistry in Context* examines a set of scientific and technological topics with broader societal implications; this manual provides laboratory experiences that are relevant to those topics.

As authors and instructors, we believe that laboratory work ought to be an integral part of chemistry courses. Hands-on experience with experimentation and data collection are crucial to an understanding of the scientific method and to the role that science plays in addressing societal issues.

The experiments in this laboratory manual are relatively straightforward with easy-to-follow instructions. Some are adaptations of traditional experiments; others are quite novel. Many use small-scale equipment of a type that has become popular in recent years. The experiments require relatively little mastery of traditional laboratory techniques; thus, maximum student time can be devoted to explorations and acquiring data.

Some of the experiments can be completed in an hour; a few require more time. For courses with a 2-hour or 3-hour lab period, it may be possible to carry out two experiments in one period or to utilize the suggestions included with many experiments for optional extensions or alternate assignments. For courses that do not have a scheduled lab component, it may be possible to do some of the experiments in a classroom setting with fairly simple modifications, especially those experiments using small-scale equipment.

The changes for the seventh edition are primarily small modifications of experiments in earlier editions. The procedures of several of the experiments have been altered for easier implementation. The introductory materials of some experiments have been modified to better emphasize the close connections between the experiments and topics in the textbook, in particular to highlight the emphasis on issues of sustainability. Five experiments from earlier editions of this lab manual are available at the *Chemistry in Context* website where they can be downloaded for class use if desired.

Instructor's notes for each experiment are available online, providing valuable information about the experiments in this manual. Instructors and lab support personnel are strongly urged to make use of this resource. The guide includes lists of materials, pedagogical objectives, teaching hints, trouble shooting, possible extensions, sample student results, and answers to questions.

This collection of experiments is the result of a collaborative effort involving many people. Major contributions to earlier editions came from Catherine Middlecamp, Norbert Pienta, Truman Schwartz, Robert Silberman, Conrad Stanitski, Gail Steehler, and Wilmer Stratton.

All of the experiments in this lab manual have been used by *Chemistry in Context* authors (past and present) in our own classes. They are versions that we have found to fit our needs. But each course is different, and instructors have differing styles. We invite users to contact us with suggestions for modifications of these experiments and/or for exchange of ideas about possible new experiments.

To the Student

The experiments in this laboratory manual have been carefully chosen and designed to reflect and amplify the contents of *Chemistry in Context.* Our goal in writing this manual was not to train you as a future chemist, but rather to show you what chemists do and how they do it. We also would like for you to appreciate the fact that there is no great mystery to doing chemistry in the laboratory. You can obtain a great deal of information about the world around you with simple chemical equipment and straightforward procedures.

Another goal is to give you an opportunity to discover how chemists solve problems. After all, chemistry is an experimental science, and most chemists spend part of their time in a laboratory. Like practicing chemists, you will use the laboratory to try out new ideas, investigate the properties of materials and compounds, synthesize compounds, analyze materials, and, in general, solve problems. Experiments are the way that scientists answer questions. Thus, you will find that each of the experiments in this lab manual begins by posing a question. The experiments show you how we try to answer those questions. Two of the experiments, *Hot Stuff: An Energy Conservation Problem* and *Solubilities: An Investigation*, ask you to solve problems as a chemist might when no instructions are available. In addition, various optional challenges or extensions are scattered throughout the manual.

You will find that most of these experiments use simple equipment and that you can easily learn the necessary techniques. In general, you will be working with a partner, and in some cases, the whole class will work collaboratively to collect data and answer a scientific question.

For nearly every experiment, you will find one or more data sheets at the end. Unless your instructor directs you otherwise, you should record all observations and data on these sheets. *It is important to make these records while you are doing the experiment, not at some later time when you may have forgotten the details.* Each experiment also includes a set of questions to be answered after completing the experiment. These questions are designed to help you consolidate what you have learned and to demonstrate your understanding.

Although you may never again work in a chemistry laboratory, it is our hope that after this laboratory course, you will understand why chemists find laboratory work so interesting and compelling.

Notes About Laboratory Safety

Good laboratory practice requires that you take some simple safety precautions whenever you work in a chemistry laboratory. The popular notion that a chemistry lab is a dangerous place, filled with unknown disasters waiting to occur, is simply untrue for most situations and is certainly incorrect for the activities in this laboratory manual. Nevertheless, all chemistry laboratories have some hazards associated with chemical spills, careless handling of flammable substances, and broken glassware. With these in mind, we now offer basic rules to follow when working in a chemistry laboratory.

1. **Wear approved eye protection at all times**. This is essential and will be rigidly enforced. Chemical splashes can harm your eyes. Even dilute solutions of many chemicals can cause serious eye damage. Fortunately, you can protect by wearing approved safety glasses or goggles. Even when you are not working directly with chemicals, someone near you may have an accident and something may splash or fly in your direction.

2. **Exercise special care when using flammable substances.** No open flames should be anywhere in the vicinity of your work. Even a <u>hot</u> <u>object</u> can sometimes cause flammable vapors to ignite. Know which liquids are flammable.

3. **Never eat, drink, or smoke in the laboratory**. You may inadvertently ingest hazardous chemicals.

4. **Never perform unauthorized experiments**. Some simple chemicals can form explosive or toxic products when mixed in unintended or inappropriate ways.

5. **Never work in a laboratory without proper supervision.** One of your best safety precautions is to have a knowledgeable person present who can spot potential hazards and handle an emergency should it arrive. Notify your instructor immediately in case of any spill or injury, no matter how small.

6. **Handle glassware carefully.** Glassware can break and can cause nasty cuts.

7. **Learn the location of fire extinguishers, a fire blanket, first-aid kit, eyewashes, and safety showers in your lab.** Be sure you know how and when they are to be used.

For many of the experiments in this text, you will find specific safety notes highlighted in the instructions. These must, of course, be followed. Develop a habit of being safety-conscious whenever you are in a laboratory.

What am I Breathing?
Preparation and Properties of O_2 and CO_2

INTRODUCTION

In Chapter 1 of *Chemistry in Context*, we ask you to take a breath (see Consider This 1.2). We will begin our laboratory investigations by examining the most important gases in that breath: oxygen and carbon dioxide. Oxygen (O_2) is necessary for life and makes up 21% of the air by volume in an inhaled breath and 16% of the air in an exhaled breath. Part of the oxygen you inhale combines with carbon compounds in your body to produce carbon dioxide (CO_2), which is then exhaled. Carbon dioxide in the atmosphere also comes from the combustion of petroleum and other fuels, and you will see elsewhere in your textbook the environmental consequences of rising CO_2 levels. In this experiment, you will prepare samples of oxygen and carbon dioxide, and then investigate some of their chemical properties.

Background Information

To prepare oxygen, you will use a catalyst to decompose a familiar household product, hydrogen peroxide (H_2O_2). A **catalyst** is a substance that participates in a chemical reaction and influences its speed without undergoing permanent change. If you have used hydrogen peroxide to clean a cut or scrape, you have probably seen an example of the catalytic reaction. Hydrogen peroxide bubbles as it is applied to the cut because a catalytic enzyme in your blood causes the peroxide to decompose and produce oxygen gas, which then helps to cleanse the cut. In this experiment, you will use potassium iodide (KI) as the catalyst to decompose hydrogen peroxide into water and oxygen.

$$2 H_2O_2 \xrightarrow{\text{catalyst}} 2 H_2O + O_2$$

Carbon dioxide will be prepared from another common household product, baking soda, which has the chemical name sodium bicarbonate ($NaHCO_3$). When hydrochloric acid (HCl) is mixed with sodium bicarbonate, they undergo a chemical reaction to form sodium chloride (NaCl), which is table salt, water (H_2O), and carbon dioxide (CO_2).

$$NaHCO_3 + HCl \longrightarrow NaCl + H_2O + CO_2$$

Both gases will be generated in "zipper" plastic bags, and you will have an opportunity to make observations about the reactions. Samples of the gases will be tested for flammability and for reactivity with a water solution of calcium hydroxide, $Ca(OH)_2$, also known as "limewater."

Finally, you will investigate what happens when these gases dissolve in water. In particular, you will determine whether or not they react with water to form an acid. Chapter 6 in your textbook explains concepts of acidity and pH as well as some of the effects that rising atmospheric CO_2 levels are having on the oceans. To test for changes in acidity, you will use an acid-sensitive dye called an **indicator** that changes color with pH. The indicator in this case, bromothymol blue, is blue in the absence of acid and yellow in the presence of acids. Thus, if a gas mixes with a water solution of bromothymol blue and the color changes from blue to yellow, this is evidence that the gas has reacted with the water to produce an acid. Air, oxygen, carbon dioxide, and exhaled air will be tested with bromothymol blue solution.

Overview of the Experiment

1. Prepare samples of exhaled air, carbon dioxide and oxygen in plastic "zipper" bags.
2. Test these gases (plus air and exhaled air) for reaction with limewater.
3. Test these gases (plus air and exhaled air) for acidity in water solution.
4. Test these gases (plus air and exhaled air) with a glowing splint.
5. Clean up.

 STOP! Safety glasses must be worn *at all times* while doing chemistry experiments.

Pre-lab Questions

1. Besides carbon dioxide and oxygen, what other gases are present in air?
2. Define *corrosive*.

EXPERIMENTAL PROCEDURE

I. Generating the Gases

General Instructions: The gases will be generated in plastic zipper-style reclosable bags, and you need to be able to seal and unseal the bags quickly and easily. It is important to <u>completely</u> seal the bags. Before proceeding further, test to be sure that a sealed bag with air in it does not leak when you squeeze it gently.

A. Exhaled Air

1. Breathe in and then exhale into a 1 pint heavy-duty "zipper" bag, repeating this 2 or 3 times; then seal the bag when it is inflated with your breath. Or: Breath in and exhale through a straw that is inserted into a 1 pint heavy duty "zipper" bag. Crimp the straw, and repeat as needed until the bag is inflated with your breath.

B. Carbon Dioxide

1. Place a teaspoonful (about 2 grams) of sodium bicarbonate, $NaHCO_3$, in the bottom corner of a 1-pint heavy-duty "zipper" bag.

2. Fill a plastic transfer pipet with 20% hydrochloric acid, HCl. To fill the pipet, try to squeeze nearly all the air out and then let the bulb fill with liquid. A second squeeze may help in getting more of the air out.

CAUTION!   HCl is corrosive and must be handled with care.

3. Place the pipet with the hydrochloric acid into the plastic bag with the NaHCO₃. Smooth out the bag so it contains a minimum amount of air; then seal the bag. (Take care not to press against the pipet.)

4. Hold the sealed plastic bag as shown in *Figure 1.1* and slowly squeeze the pipet so that the acid drops onto the NaHCO₃. Observe carefully what happens (several changes should be apparent) and record your observations on the data sheet. **Keep the bag sealed.**

5. You should now have a sealed plastic bag partially filled with carbon dioxide (CO_2) gas. *Leave the bag standing upright (zipper on top) by leaning it against something on the lab bench.*

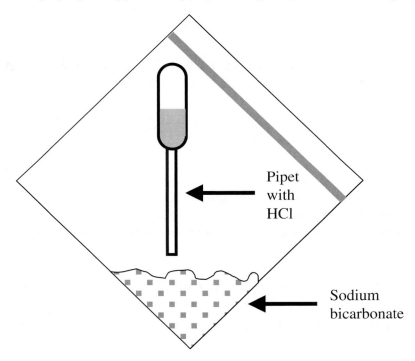

Figure 1.1 Sodium bicarbonate in the "zipper" bag

C. Oxygen

1. Use a spatula to place a small pinch of potassium iodide (0.5 gram or less) in the bottom corner of another 1-pint heavy-duty "zipper" bag.

2. Fill one or two plastic transfer pipets with 10% hydrogen peroxide (H_2O_2). See directions above for filling the pipets.

 CAUTION!   Hydrogen peroxide is corrosive and must be handled with care.

3. Place the pipet(s) with 10% hydrogen peroxide into the plastic bag with the potassium iodide. Smooth out the bag so it contains a minimum amount of air and then seal the bag. (Take care not to press against the pipet.)

4. Hold the sealed plastic bag as shown in *Figure 1.1* and slowly squeeze one pipet so that the H_2O_2 drops onto the potassium iodide. If a second pipet was used, wait a few moments and then gently squeeze it also. (Try to get all of the liquid *out* of the pipets.) Observe what happens and record your observations on the data sheet. **Keep the bag sealed.**

5. You should now have a sealed plastic bag partially filled with oxygen (O_2) gas. *Leave the bag upright (zipper on top) by leaning it against something on the lab bench.*

D. Room Air

1. Squeeze room air in and out of a clean pipette and then use this captured room air in Part II.

II. Properties of Air, Carbon Dioxide, Oxygen, and Exhaled Air

General instructions for the tests: Use a plastic transfer pipet-full of gas to perform each test on each of the gases you generated in this experiment. Record all of your observations on the data sheet. Use a fresh pipet (clean and dry) for each test. It helps to have two students working together to fill the pipets by the following method.

To get a pipet-full of a particular gas, squeeze the bulb to expel as much as possible of the air inside the pipet. Keep squeezing the bulb and slowly push the tip of the pipet against the zip seal at one corner of the plastic bag containing the gas to be tested. With a bit of practice, you will be able to just push the pipet tip so that the seal opens around it. Taking care not to touch the liquid or solid chemicals in the bag with the pipet, push the pipet tip into the bag. Then release the bulb so that gas enters the pipet. Quickly withdraw the pipet. As the tip leaves the bag, immediately reseal the bag along the "zip strip."

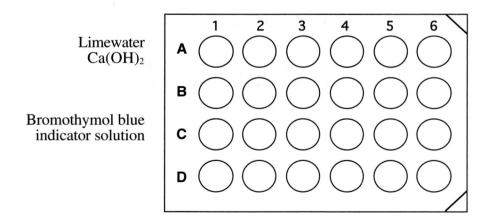

Figure 1.2 The reaction wellplate setup

A. Limewater Tests

1. In row A of a wellplate (*Figure 1.2*), add 10 drops of limewater to wells A1, A2, A3, and A4. Place the wellplate on a dark surface or a small piece of black paper.

2. Fill a clean pipet with carbon dioxide gas. Place the tip of the pipet into well A1. By gently squeezing the pipet, <u>slowly</u> bubble the gas through the limewater solution. Observe and record the results on the data sheet.

3. Repeat the test using a sample of oxygen gas bubbled into well A2. Record your observations.

4. Repeat the test using a sample of exhaled air bubbled into well A3. Record your observations.

5. Repeat the test using a sample of ordinary air bubbled into well A4. Record your observations.

B. Tests with an Indicator Solution

1. In row C of the wellplate, add 10 drops of bromothymol blue indicator solution to wells C1, C2, C3, and C4. Place the wellplate on a light-colored surface or a piece of white paper.

2. Fill a clean pipet with carbon dioxide gas. Place the tip of the pipet into well C1. By gently squeezing the pipet, <u>slowly</u> bubble the gas through the indicator solution. If the indicator changes color to yellow, it indicates that the solution has become acidic. Record the results on the data sheet.

3. Repeat the test using a sample of oxygen gas bubbled into well C2. Record your observations.

4. Repeat the test using a sample of exhaled air bubbled into well C3. Record your observations.

5. Repeat the test using a sample of ordinary air bubbled into well C4. Record your observations.

C. Glowing Wood-Splint Test

1. Use a clean, dry pipet to obtain another sample of carbon dioxide gas.

2. With a match, ignite the end of a wood splint or toothpick (or simply use a wood match). After it has burned for a few seconds, blow out the flame. Continue to blow on the embers so that they glow.

3. Have your lab partner hold the pipet filled with carbon dioxide so that the tip is very near the glowing ember and *gently* squeeze a puff of carbon dioxide gas directly at the glowing portion. Observe and record the results.

4. Using the same pipet, repeat this test using a sample of oxygen.

5. Finally, repeat this test using samples of exhaled air and ordinary air.

Clean-up

Rinse the plastic pipets with water, and clean the wellplate with soap and a brush. Discard the solutions and the bags in appropriate waste containers as your instructor directs you. Do not put *anything* down the sink drain unless you are told that is permitted.

Post-lab Questions

1. What role does potassium iodide play in the generation of oxygen? How does this differ from the role of sodium bicarbonate in the production of carbon dioxide?

2. What evidence from today's experiment supports the idea that exhaled air contains more CO_2 than room air?

3. According to Chapter 6 of *Chemistry in Context*, the oceans absorb 25-40% of all anthropogenic CO_2 emissions.
   a. Based on your observations in this experiment, what effect will this have on the pH of the oceans?
   b. List two consequences of ocean acidification. (Hint: see section 6.5 of your text.)

4. Based on what you observed about the interaction of carbon dioxide with the glowing splint, explain how CO_2 fire extinguishers work.

5. Based on what you observed about the interaction of oxygen with the glowing splint, explain why liquid oxygen is an extremely hazardous material.

6. If your blood becomes too acidic, you may begin to hyperventilate, which causes your blood O_2 levels to increase, and your CO_2 levels to decrease. Write the equation for reaction of CO_2 with water, and explain why reducing CO_2 in the blood will reduce the acidity of the blood.

7. Propose a way to estimate the volume of oxygen that was generated in the plastic bag.

8. **Optional challenge:** Propose a way to determine the amount of carbon dioxide in a carbonated beverage such as a club soda, beer, or cola drink.

9. **Optional challenge:** What is the purpose of including room air as a standard for the limewater and bromothymol blue indicator tests?

Is it Pure or a Mixture?
Chromatographic Study of Dyes and Inks

INTRODUCTION

How can we know if something is pure, containing just one element or compound, or a mixture of several substances? Section 1.6 in *Chemistry in Context* introduces the concepts of pure substances and mixtures, and notes that most of the materials we encounter daily, such as air, water, and food, are mixtures of chemical substances. Chromatography is a common method for separating and studying the components of a mixture. This experiment uses paper chromatography to identify whether certain colored dyes and inks are pure substances or mixtures.

Chromatography is the study of separations of mixtures and is often used to identify unknown components in mixtures. In chromatography, the components of a mixture are allowed to move along a stationary surface such as a sheet of paper. Each component in a mixture retains its own properties and thus moves at a rate determined by its own characteristics. Since they move at different rates, the components become spread out and separated from each other like runners in a foot race. When paper is the stationary surface, the usual strategy is to place small spots of substances near one edge of the paper. That edge is placed in a liquid such as water, allowing the liquid to move up the paper by capillary action. The liquid carries the components with it, but at differing rates depending on their solubility in water.

Background Information

The word "chromatography" derives from the Greek words for "color" and "write". Historically, the technique first utilized colors to see and identify the components of a mixture (much as you will do in this study). Today, a variety of other methods are employed to observe the separated components of mixtures. (Experiment 24 in this laboratory manual utilizes a slightly different chromatographic method to identify analgesic compounds.)

In this brief exploration, you will investigate artificial dyes used in drink mixes and/or several felt-tip pens of the type used with overhead projectors to find out whether each color is produced by a pure substance or a mixture of substances. By careful comparison, it should be possible to determine whether some of the substances have colored components in common.

Overview of the Experiment

1. Place spots of colored dyes or inks on a piece of filter paper.
2. Immerse one edge of filter paper in water and allow water to move the spots upward.
3. Compare the spot locations for the various inks.

Pre-lab Question

In your own words, suggest some reasons why the dyes will move up the filter paper at different rates. Some things to consider are chemical structure, solubility, and affinity of some substances for others.

EXPERIMENTAL PROCEDURE

These experiments should be done individually, so each person has a completed chromatogram, but you are encouraged to compare results and discuss the interpretation. You instructor will tell you which procedure to follow, or you may do both.

A. Food dyes

1. Obtain four different drink mixes (such as Kool-aid).

2. If solutions of the drink mixes have not already been prepared, you will need to make them yourselves. Using a balance, weigh 0.5 g of each drink powder, placing each mix in its own test tube. You instructor will explain the correct use of the balance. To each test tube, add 6 drops of water. Shake the test tube and continue to add water one drop at a time until the powder is just dissolved. It is important to have a concentrated solution.

 STOP! Do not drink any beverages that have been opened or used in the laboratory.

3. Obtain a rectangular piece of filter paper, approximately 5 cm x 12 cm. Using a *pencil*, write your name at the top and then draw a line 2 cm from the bottom edge of the paper, as shown in *Figure 2.1*. Make four small pencil marks 1 cm apart along the line you just drew.

4. Using a small capillary tube, place a spot of one of your solutions on the first mark. Keep the spot small. You may need to spot in the same place two or three times, allowing the spot to dry before spotting again.

5. Make spots with the other solutions on the other marks, using a new capillary tube for each solution. In pencil, label the name of the solution spotted on the mark.

6. Obtain a 400 mL beaker and a glass rod to lay across the top of the beaker. Fold or roll the top edge of the filter paper so that it hangs on the glass rod inside the beaker, but the bottom of the filter paper does not touch the bottom of the beaker, as shown in *Figure 2.1*.

7. Carefully add a small amount of water so that there is a shallow pool of pure water in the beaker with the bottom of the paper extending into it. Leave the beaker undisturbed on the desk and observe what happens.

8. When the top edge of the water has moved about 2/3 of the way up the paper, remove it and lay the paper out on a paper towel to dry. While still damp, draw a line *in pencil* to show the top edge of the wet area.

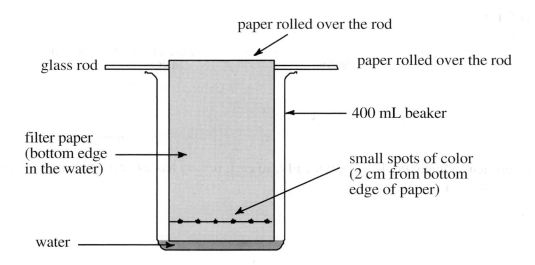

Figure 2.1 Experimental set-up for chromatography experiment.

B. Water soluble inks

1. Obtain four different-colored pens that contain vivid, washable inks. Good choices are the Vis-à-Vis overhead-projector pens or LiquidMark Washable Markers.

2. Follow steps 3 through 8 above. However, rather than using a capillary tube to make your spots, you can use the pens directly to draw a small spot on the filter paper.

Optional Variations or Extensions

1. Investigate what happens with other kinds of pens. Some possibilities: other transparency pens, "Sharpie" or "Flair" pens, and ballpoint pens. (For ballpoint pens, make sure you have a good dark spot.) Use a fresh sheet of filter paper for each investigation. For permanent markers or ballpoint pens, try replacing the water with a solution of alcohol or acetone in water.

2. Investigate what happens if you replace the water with some other solvent, such as methyl alcohol, ethyl alcohol, or rubbing alcohol (which is a mixture of 70% isopropyl alcohol and 30% water).

3. Using a circular piece of filter paper, cut a 1 cm path to the center of the paper disk. Spot six colors about 1 cm from the center of the circle. Fold the paper tab down and place the filter paper on the rim of a 50 mL beaker filled with enough water that it comes into contact with the tab. You will see the colors fan out towards the edge of the filter paper in a floral pattern.

4. Try doing the experiment with a natural substance, such as pigments from plants or flowers. You can extract the pigments from the leaves or petals of the plants by grinding in a mortar with rubbing alcohol, and then filtering the resulting solution. Spot the filter paper using a capillary tube as described above.

Report

When the filter papers are completely dry, staple or tape each one to a sheet of 8 1/2 x 11 inch paper. Label the sheet with your name and lab section and include a brief summary of what samples and solvent were used. On a separate sheet, answer the questions on the next page.

Red   purple   blue   orange

Kool-aid
K#1   K#2   K#3   K#4

Post-lab Questions

1. Which of the dyes or inks appear to contain a single colored substance, and which are mixtures? Describe your evidence.

2. Which samples have colored components in common with other samples? Explain your reasoning.

3. From your results, predict which colored components are most soluble in water and which are least soluble in water. Explain your reasoning.

4. Describe an experiment you could do in your kitchen to convince a friend that the hard coating on an M&M contains a mixture of several food colorings. Explain what you would do and what you would see as results.

5. (If you did optional step 1) An interesting household problem is that of removing ballpoint pen stains from clothing. In light of your observations, can you suggest what will and will not likely work? What additional tests could you do to investigate this problem further?

6. (If you did optional step 2) Suggest a possible explanation for why the results using alcohol are different from those in water.

Can I Spot a Trend?
A Graphic Experience Weighing Air and Cooling Water

INTRODUCTION

Graphs provide an important and very useful way to present data. They efficiently summarize numerical data and are usually easier to understand and interpret than columns of numbers. In this assignment, you will collect some experimental data that lends itself to presentation in graphical form. You will then learn how to construct graphs in ways to make the visual presentations most effective.

Two short laboratory exercises are included. Your instructor will specify whether you are to do one or both, depending on the available time. The first exercise is a study of the relationship between pressure and mass of air in an empty soft-drink bottle. The second investigates the relationship between time and temperature as a hot liquid cools.

Background Information

Graphs are essentially a pictorial way of presenting information. For example, *Figure 3.1* presents some data on the relationship between mass and volume for a certain substance. The data is presented in both tabular and graphical form. It can be seen that there is a direct relationship between the mass and the volume for this substance.

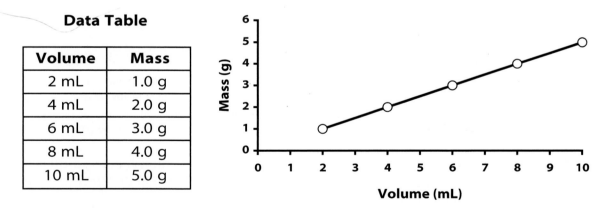

Data Table

Volume	Mass
2 mL	1.0 g
4 mL	2.0 g
6 mL	3.0 g
8 mL	4.0 g
10 mL	5.0 g

Figure 3.1 Graphical representation of data: volume as a function of mass

The graph in *Figure 3.1* shows that as the volume increases, the mass increases in direct proportion. Data that produce a straight line when plotted is said to have a linear relationship. Representing data with a graph makes it straightforward to estimate the value of the mass for a volume that is in-between measured data points, a process known as **interpolation.** As an example, if the volume were 5 mL, it is easy to see that the corresponding mass would be 2.5 g. It is also easy to extend the line of the graph and obtain data beyond the range of measured points, a process known as **extrapolation.** Thus, for example, if the volume were 15 mL, you should be able to convince yourself that the mass would be about 7.5 g. Care must be taken when extrapolating certain types of data, however because the trends observed in one range of a graph are not necessarily sustained in other regions of the same graph.

To make graphs easy to read and interpret, the following conventions should be observed.

1. Graphs should be neat, legible, and well organized.

2. There should be a descriptive title designed to tell the reader what has been plotted. An example might be: "volume of plastic as a function of its mass." Titles such as "M vs. V" are too cryptic and should not be used.

3. The horizontal axis (*x* axis) and the vertical axis (*y* axis) should be clearly labeled to show what is plotted (e.g., volume) and the units (e.g., mL).

4. The *scale* of the graph should be chosen so that the graph fills as much of the paper as practical. In general, the scales on the *x* axis and the *y* axis do not need to start at 0. *Figures 3.1* and *3.3* show examples of correctly scaled graphs. The data points fill the graph space quite effectively. *Figure 3.2* shows three examples of incorrectly scaled graphs.

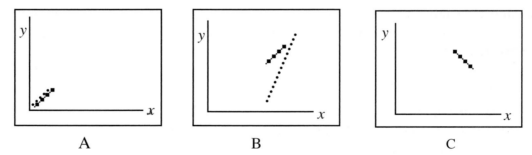

A   B   C

Figure 3.2 Examples of incorrectly scaled graphs

Graph A: The increments on the *x* and *y* axes should be made smaller to spread the graph out over the page.

Graph B: The starting point on the *x* axis should be changed and the increments on the *x* axis should be made smaller so the graph takes up most of the page.

Graph C: The starting point on both the *x* and *y* axes should be changed and the increments on both axes should be made smaller so the graph takes up most of the page.

5. If the data points appear to represent a linear (straight-line) relationship, use a ruler to draw a single straight line that best represents the average relationship. On the other hand, if the data points seem to follow a curve rather than a straight line, then draw the best *smooth* curve through them. (A curve is harder to draw than a straight line.) It is unlikely that all of the points will fit exactly on one smooth line (either straight or curved); therefore, some judgment must be exercised in deciding on the "best" fit. But it is not good form in science to draw zig-zag "connect-the-dot" lines.

6. When some sets of experimental data are plotted, it becomes apparent that certain points seem "out of line" with all of the others. (Scientists often call these "outliers.") In this case, it is quite likely that some error was made, either in the original measurement or in writing it down. (It is also possible that an error was made in placing the point on the graph—this is the first thing to check.) If a point really seems to be out of line with the others, it is appropriate to exclude it when deciding on the best line. Clearly, this requires some judgment.

A straight-line relationship such as the one in *Figure 3.1* can be summarized with a simple algebraic equation of the form $y = mx + b$. In this equation, x and y are the values for the two quantities being plotted (e.g., volume and mass), m is the **slope** of the plotted line, and b is the **y-intercept** (the value of y when the line crosses the y axis or, in other words, the point where $x = 0$). The **slope** of a line is calculated by determining the difference between the y values for two data points and the difference between the x values for the same two data points. The equation for this is

$$\text{slope} = m = \frac{y_2 - y_1}{x_2 - x_1}$$

The slope summarizes the relationship between the columns of data. In the example above, it is easy to see that the slope is 0.5, and it represents the relationship between mass and volume, or the density of the substance. Because the line crosses the y axis at 0, the y-intercept is 0, and the equation that represents the line is $y = 0.5x + 0$ or simply $y = 0.5x$. A straight-line relationship of the type shown in the figure is the easiest to construct and the easiest to interpret; therefore, researchers often go to great lengths to discover a linear relationship when a plot of raw data does not yield one directly.

Computer Aids to Graphing

Computer programs are available to take much of the guesswork and tedium out of constructing graphs. Many programs, for example, can create a complete graph from your data and provide you with a printed copy of the graph. Most of these programs can do what is called a "linear regression analysis," which uses a statistical method to calculate the best straight-line fit for a set of data points and then to draw the line. This can be particularly useful when there is a lot of "scatter" in the data, as in the example shown in *Figure 3.3*. Consult your instructor to determine if a computer-graphing option is available. Even if computer assistance is available, it is still useful to learn how to construct graphs on your own. And even with computer programs, some judgment is needed about choice of scales and whether any "outlier" points should be excluded.

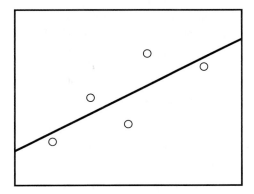

Figure 3.3 Linear regression fit of data points

Overview of the Experiment

PART I
1. Obtain an empty 2-liter bottle with a tire valve mounted in the cap.
2. Inflate the bottle to about 40 pounds per square inch with a bicycle pump.
3. Weigh the bottle and measure the pressure in the bottle.
4. Release some of the air; measure the pressure and weigh the bottle again.
5. Repeat step 4 several times until all of the air pressure has been released.

PART II
1. Heat some water in a beaker to a temperature of about 80°C.
2. Allow the water to cool and record its temperature at regular time intervals.

 STOP! Safety glasses must be worn *at all times* while doing chemistry experiments.

Pre-lab Questions
1. Air has a mass because it is composed of chemical compounds. What are the compounds that you'll be weighing today?
2. Do you expect the graphs you'll construct to have linear or curved regressions? Why?

EXPERIMENTAL PROCEDURE

I. Weighing Air

1. Before beginning the measurements, you need to learn how to use a laboratory balance for measuring the mass of an object. Your instructor will explain the use of the particular balances in your laboratory. If you are using an electronic-type balance, it is important to make sure the balance reads zero when empty (usually accomplished by pressing a TARE button). It is a good idea to recheck the zero each time you use the balance. For this experiment, you will need to make measurements to the nearest 1/100 of a gram (0.01 gram).

2. Obtain a clean, dry, plastic soda bottle (2-liter size preferred) with a screw cap that has been fitted with a tire valve. Put a thermometer strip into the bottle and tightly screw on the cap.

3. Record the temperature of the air inside the bottle. Each time you manipulate the bottle, record the temperature from the inner thermometer. Also record if the bottle feels warm or cool to the touch, and any other observations.

4. Become familiar with using the tire gauge to measure the air pressure. Ask your instructor for assistance.   You will need to be able to push the gauge on squarely without letting out much air, and to know how to read the pressure units on the protruding stem. The pressure scale is probably marked in pounds-per-square-inch ("psi").

5. Attach the hose from a hand tire pump to the valve. Place the bottle in a protective enclosure (such as a wastebasket). Pump up the pressure in the bottle to about 40 psi, as measured by a tire gauge.

6. Weigh the bottle and record the mass. Wait 5 minutes and reweigh the bottle.

7. If the mass of the bottle (plus air) is the same as it was initially, the cap is sealed tightly and the experiment can be started. If the bottle's mass decreased by more than 0.05 g, tighten the cap and repeat steps 4 and 5.

8. If necessary, pump up the pressure again to 30–40 psi. Carefully measure the air pressure in the bottle using the tire gauge and immediately record this in the data table. Then weigh the bottle on a balance and record the mass (to the nearest 0.01 gram) and temperature.

9. Let a small amount of air out of the bottle. Measure and record the pressure and temperature again; then re-weigh and record the mass.

10. Repeat step 8 until you have obtained at least five sets of pressures, temperatures and masses between your highest reading and 10 psi.

11. Finally, let all of the air out, reweigh and record the temperature. (Of course the bottle is not really empty. It now has the same air pressure as the surrounding air, about 14.7 psi.)

12. **OPTIONAL:** These measurements can be done very rapidly. If time permits and you are interested, you might like to try repeating the study. (In scientific studies, the measurement technique often improves with repeated experiments.)

II. Cooling Water

 STOP! This part of the experiment involves an open flame. No flammable chemicals should be in the vicinity. Long hair should be tied back, and extremely loose sleeves on clothing should be avoided.

1. Use a thermometer or temperature probe to measure the air temperature in the room. Record this on the data sheet.

2. Measure 50 mL of water using a graduated cylinder, pour it into a 100 mL beaker, and place the beaker on a ring stand over a Bunsen burner.

3. Light the Bunsen burner and heat the water to about 80°C. Turn the Bunsen burner off.

4. Without stirring or removing the beaker from the ring stand, allow the water to cool to 75°C.

5. When the water has cooled to 75°C, start recording the temperature at intervals of 2 minutes. Use a wall clock, wrist watch, or stop watch to keep track of the time. Record the temperatures to the nearest 0.5 degree.

6. Continue making measurements until the water has cooled to about 35°C.

III. Constructing Graphs of Your Data

Part I

1. Look carefully at the range of recorded masses from lowest to highest. For the vertical (*y*) axis on the graph, select a *convenient* set of units that will include this range of masses. (It helps to select units with subdivisions of 10 since the masses are in decimal units.)

2. Similarly, look at the recorded pressures. (They should range from 0 to about 30 or 40 psi.) Choose a convenient set of units for the horizontal (*x*) axis.

3. Carefully plot the experimental points. (Use a pencil because it is easy to make mistakes in this kind of plotting.) Make small dots, and then draw small circles around them so that they show up clearly.

4. Examine the data points carefully. Using a ruler or other straight edge (preferably transparent), draw the "best" straight line through the data points. Don't do this too hurriedly—it requires some judgment. If the points are somewhat scattered, there should be equal numbers of them on either side of the line.

5. An interesting further use of this data set is to extrapolate the line to −14.7 psi. (If a computer-graphing program is available, this becomes easy to do simply by changing the scale on the horizontal axis.) What do you think you will see?

6. If specified by your instructor, calculate the slope of the line. To do this accurately, DO **NOT** use particular data points. Instead, select two places on the line near the opposite ends of the line. Carefully read off the *x* and *y* values for each, and then use the equation given in the Background section to calculate the slope.

Part II

1. Select a convenient set of units on the horizontal (*x*) axis, starting at zero, to cover the number of minutes over which you made measurements.

2. Select a convenient set of units on the vertical (*y*) axis to cover the temperature range 35 to 75°C.

3. Proceed as in Part I to plot the points. Use pencil.

4. It is unlikely that these values will fall on a straight line. Draw the best curve you can through the points. (Your instructor may offer some helpful tips for drawing a smooth curve.)

Post-lab Questions

Part I

1. Suppose you could pump some air out of the bottle rather than pumping it in. Predict how the mass would change. Is there a limit to the change?

2. All of your mass measurements were actually the combined mass of the bottle itself plus the air it contained. How could you find out the mass of the air alone? (Hint: What is the significance of the weight of the bottle at −14.7 psi?)

3. Suggest a possible explanation as to why the bottle warmed and cooled when it did.

Part II

4. Write a paragraph describing the advantages of using a graph to represent data rather than just using a data table. Include examples encountered in this experiment's introduction and results.

5. Figure 3.3 in your textbook shows two plots of well-known data describing the increase in atmospheric carbon dioxide with respect to time. Study this figure and answer the following questions.

    a. What two quantities are plotted against each other in each graph? What are the units?

    b. The data from Mauna Loa is included in the plot spanning thousands of years. Why is it useful to have this data plotted separately, as in the inset figure?

    c. People continue to argue about the validity of extrapolating this data to predict future atmospheric carbon dioxide levels. In what ways might an extrapolation be valid, and what concerns may be raised by trying to extrapolate this data?

6. Choose another graph from your textbook, write down the figure number, and answer the following questions.

    a. What quantities are plotted against each other in the graph? What are the units?

    b. Describe what this graph tells you about the relationship between these quantities (i.e. does one quantity increase or decrease as the other increases?). Does the graph indicate correlation, causation, or no relationship between these quantities? (Look up *correlation* and *causation* if you don't know what these terms mean.)

Notes

What Protects Us from Ultraviolet Light?

INTRODUCTION

Ultraviolet light has sufficient energy to cause changes in DNA and thus skin cancers. The dangers of exposure to ultraviolet light are described in Section 2.7 of *Chemistry in Context*. In recent years, depletion of the ozone layer has allowed more ultraviolet light to reach us, resulting in more cases of skin cancer. Consequently, we have become increasingly aware of the need to protect ourselves from exposure to ultraviolet light.

One strategy to avoid exposure to ultraviolet light would be to avoid exposure to any type of light. Few among us would be willing to take this to the extreme of moving underground, but we've all moved under a shady tree or used a hat to block out light.

A wide variety of commercial lotions are available with sun protection factors (SPF) ranging from 2 to over 50. These lotions contain organic molecules such as oxybenzone and octyl methoxycinnamate that absorb light specifically in the ultraviolet region of the spectrum. Similarly, we can purchase sunglasses whose lenses absorb either UV-A (320–400 nm) or UV-B (280–320 nm) light, or both. Throughout the years, sunbathers have tried any number of homespun methods to protect themselves: slathering on baby oil, spraying mists of water, or sunbathing behind a glass or plastic panel.

The Problem

Your assignment is to design and conduct an experiment to test the ability of a minimum of three materials to protect us from ultraviolet light. You will be given a handful of beads that change color upon exposure to UV light. The more UV light they are exposed to, the faster their color will change. Thus, by measuring how long it takes the beads to change color, you are measuring the amount of UV light reaching them. You will also be provided with a variety of protective materials to test. Your instructor will give you a list of what is available and may allow you to bring additional items from home.

As you conduct your experiment, remember to protect the beads from all sides. You will need to take your experiment outdoors, but since the beads are quite sensitive, direct sunlight is not required. In fact, working in a shady spot usually gives better results. The beads will quickly change back to their original color if you put them in a dark place such as your pocket. Good experimental design requires controls. You want to change only one variable at a time. You will also want to know how the beads respond when left totally unprotected. Multiple trials should be used so that average data can be reported.

Overview of the Experiment

1. Design your experiment using a minimum of three protective materials and with appropriate controls included.
2. Get instructor approval/suggestions for your procedure.
3. Conduct your experiment.
4. Write your report.

Pre-lab Question

Form a hypothesis ranking your proposed materials in order of their ability to protect the beads from UV light. Explain the basis of your reasoning.

EXPERIMENTAL PROCEDURE

1. With your partner or other team members, develop a plan for your experiment. On your own paper, outline the procedure that you intend to follow. Be sure to include (a) three protective materials, (b) descriptions of controls, and (c) multiple trials. Specify what you will measure, how and when you will do it.

2. Obtain instructor approval to proceed. (Your instructor may offer some suggestions or ask questions to help you refine your experimental plan.)

3. Conduct your experiment. Keep a careful written record of what you do, noting especially any changes from your original plan. Record your data in an appropriate table of your own devising.

4. Prepare a written report, in which you should summarize
    a. the purpose of the experiment
    b. the procedure you followed
    c. your data in clearly labeled tables
    d. your conclusions regarding
        i. how well each material protected the beads from UV light
        ii. your controls and multiple trials
        iii. any experimental difficulties
        iv. proposals for future improvement in the experimental design

Post-lab Questions

1. Compare the energy of ultraviolet light photons to that of infrared and visible light photons. Explain why UV light is potentially more dangerous than visible or infrared light.

2. What is the purpose of an experimental control? Give one specific example of a control you used and what information it provided for you.

3. Why are multiple trials of the same procedure needed? Give one specific example of an undesirable result that could occur if multiple trials were not used.

4. Describe one specific change that you would make in your procedure if you were doing the experiment over. Explain how this change could give better or more interesting results.

5. Most sunscreen lotions and sunglasses claim to protect against UV-A and UV-B light. Why don't they mention UV-C light? Is it less dangerous than the other types of UV light? What happens to the UV-C? (See Section 2.6 and Table 2.4 of *Chemistry in Context*.)

6. On the next page, spectra of several materials that were tested for their ability to give protection from UV light are shown. Each spectrum shows the wavelength range(s) at which the material absorbs light. If a material has a large absorbance at a particular wavelength, it will protect us from light of that wavelength. If absorbance is low, the material will not provide effective protection.

   a. <u>On each of the spectra</u>, label these types-of-light regions along the *x*-axis:
      UV-A light 320–400 nm
      UV-B light 280–320 nm
      UV-C light below 280 nm

   b. Examine the spectrum of glass. Which types of light (UV-A, UV-B, and/or UV-C) does glass absorb best? From what type of UV light does this glass fail to protect us? Explain.

   c. Now look at the spectrum of clear plastic wrap. What types of light does the plastic wrap absorb? Would plastic wrap be good for preventing sunburn or skin cancer? Explain.

   d. What types of light does the sunscreen absorb?

   e. Finally, look at the spectrum of cotton fabric. How is this different from all the other spectra? Does the fabric selectively absorb certain wavelengths? What is the fabric doing that is different?

7. What materials do the best job of protecting us from UV light? Give examples from your experiment.

Ultraviolet Absorption Spectra of Protective Materials

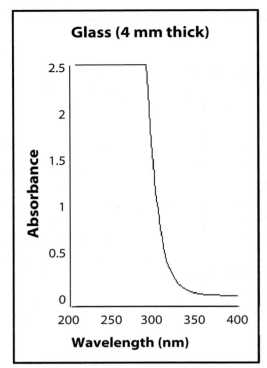

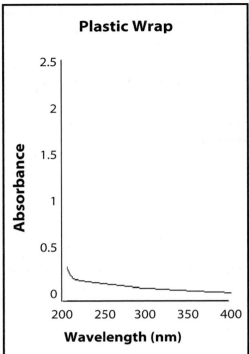

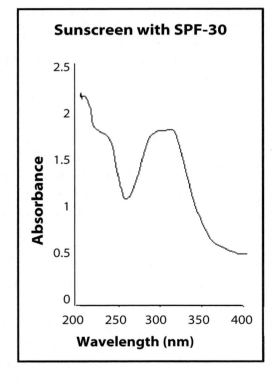

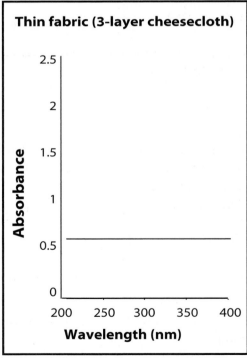

Visibly Delighted:
How Do Colored Solutions Interact with Light?

INTRODUCTION

There is something magical about rainbows and colored light from glass prisms. The separation of white light into its colored components is the actual reason for these natural phenomena. In this experiment, you will investigate this partitioning of light in the visible region of the spectrum and its interaction with colored solutions. You will discover which colors correspond to which wavelengths, and, for each of several colored solutions, you will collect the necessary data to construct a spectrum showing how light is absorbed at different wavelengths for that substance. Measurements of this kind are widely used in nearly all areas of science.

An understanding of the electromagnetic spectrum and the interactions of radiation with matter is crucial to two of the case studies in *Chemistry in Context:* the stratospheric ozone shield (Chapter 2) and global warming (Chapter 3).

Background Information

The **electromagnetic spectrum** is discussed in Section 2.4 and *Figure 2.6* of the text. The region that the human eye can detect with the wonderful rainbow of colors (the so-called "visible" region) is only a small portion of the total electromagnetic spectrum. Each color and each unique position in the electromagnetic spectrum is identified by its **wavelength**. For the visible region, the lengths of the light waves are in the range of billionths of a meter, and, therefore, the wavelengths are expressed as **nanometers**, nm (1 nm = 1 x 10^{-9} meter).

The instrument used to measure the interaction of light with matter is a **spectrophotometer**. This device is designed to split visible light into its component colors (i.e., different wavelengths) and then allow light of a selected wavelength region to pass through a sample of the material being studied. An electronic detector measures the amount of light that has been transmitted or absorbed by the sample at each wavelength. All spectrophotometers have essentially the same major components but with varying degrees of sophistication. These essential components are (1) a light source, (2) a device to isolate or resolve particular wavelengths of light, (3) a sample holder, (4) a detector, and (5) a meter or other device to display the measured transmittance or absorbance of light. *Figure 5.1* (next page) shows a simple block diagram of spectrophotometer components.

Spectrophotometers are available for most regions of the electromagnetic spectrum. Spectrophotometers for the visible region of the spectrum were developed first and are still the most common. In these instruments, the light source is an ordinary incandescent lightbulb; the wavelength selector consists of a diffraction grating and some lenses; a liquid sample is placed in a container resembling a test tube; and the detector is a phototube that converts light intensity into an electrical signal.

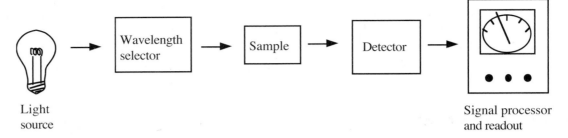

Figure 5.1 Block diagram of spectrophotometer components

Unfortunately, a simple spectrophotometer of this type can measure the transmittance of light of only a single wavelength at a time. Obtaining data over a range of wavelengths requires multiple measurements, each made at a different wavelength. Because light intensity and detector sensitivity vary with wavelength, each measurement must be corrected using a **blank**. For this experiment, the blank is simply a test tube containing pure water.

The meter on the instrument you use will display either **absorbance (A)** or **percent transmittance (%T)**. The percent transmittance unit focuses our attention on the light that passes unaffected through the sample. Percent transmittance is defined as the amount of light that passes through a sample of known thickness (the "path length") divided by the light that passes through a blank of equal thickness, and multiplied by 100.

$$\% \text{ transmittance } (\% \text{ T}) = \frac{\text{transmittance by sample}}{\text{transmittance by blank}} \times 100 \%$$

Absorbance, the other possible unit, focuses our attention instead on the portion of light that gets absorbed by the sample. Absorbance is proportional to the concentration of the solution. A sample that absorbs no light will have 100% transmission and zero absorbance. Either unit may be used in this experiment. Your instructor will choose one or the other for the entire class to use.

When you make your measurements, you should use two matched test tubes or other containers: one containing the sample and one containing pure water. With the distilled water blank in place, you will adjust the spectrophotometer to read 100% T or zero absorbance. You will then insert your sample and record data for your sample by simply reading the displayed value. From your data, you will construct a graphical representation of the light transmittance or absorbance of a colored compound over a range of wavelengths. This is accomplished by taking measurements at different wavelengths and plotting this data on the y axis versus the wavelength on the x axis. The result is a **spectrum**, a curved line that shows how a particular solution transmits or absorbs light in the visible region of the electromagnetic spectrum.

In this experiment, you will first determine the colors of different wavelengths of light throughout the visible region. Then you will determine which wavelengths of visible light are transmitted or absorbed by colored solutions. The colored liquids you will use for this part of the experiment are food-coloring dyes. Your instructor may have you do one or more optional extensions. You may be able to investigate a substance of your own choosing. This can be any colored substance so long as it meets two conditions: (a) It must be transparent so that light will pass through it, and (b) it needs to have a fairly bright, distinct color.

One category of colored substances of particular interest to chemists is acid-base "indicators," colored dyes that are used to identify the acidic or alkaline character of solutions. You have probably already used one such substance (bromthymol blue) in Experiment 1. Others will be used in later experiments in this laboratory manual. Your instructor may ask you to do a spectrophotometric study of such an indicator, either in addition to, or in place of, the study of food-coloring dyes. If so, this will entail measuring the absorption spectrum of the indicator in both its acidic and alkaline forms.

Overview of the Experiment

1. Prepare a piece of chalk to reflect light from the spectrophotometer source.
2. Determine the color of light of selected wavelengths from 400 to 700 nm.
3. Measure the transmittance (or absorbance) spectrum for a red solution and a blue solution.
4. Optional: Also measure the spectrum for a green solution.
5. Plot a graph of transmittance (or absorbance) vs. wavelength for each colored solution.
6. Use this procedure to investigate a colored substance of your own choosing.

Pre-lab Question

Do you expect the red solution to absorb light at a shorter or longer wavelength than the blue solution?

EXPERIMENTAL PROCEDURE

Note: *The following instructions assume that you are using either an analog Spectronic 20 Spectrophotometer (with knobs and dials) or its digital equivalent (with buttons and digital display). If you are using a different kind of spectrophotometer, your instructor will provide alternative instructions.*

I. Looking at Light

1. Turn on the spectrophotometer and let it warm up.

   If you have an analog Spectronic 20, investigate what the three control knobs do. The knob on top is used to set the wavelength. The left front knob is the "zero" knob, used to adjust the meter to 0% T when the light path is blocked, and the right front knob is the "transmittance" knob, used to adjust the meter to 100% T when the blank is in place.

   If you have a digital spectrophotometer, choose either the %T or Absorbance mode as directed by your instructor. Investigate how to adjust the wavelength (usually with nm-up and nm-down buttons).

2. Take a half-inch long piece of chalk and rub it on the blackboard until one end is worn down to a forty-five-degree angle (*Figure 5.2*). Place the chalk in the spectrophotometer test tube and insert it into the spectrophotometer so that the slanted side is pointing to the right (see diagram).

3. Adjust the wavelength to 500 nm (nanometers or 10^{-9} meter). If you are using an analog meter, turn the transmittance knob (front right) all the way to the right.

4. Look down the tube to the slanted piece of chalk. You should see a colored band of light. If not, raise, lower, or rotate the test tube slightly until you do. If you still cannot see the colored band, consult your instructor.

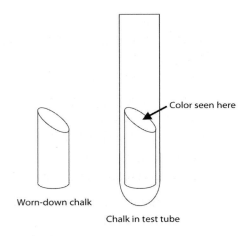

5. Now slowly adjust the wavelength in both directions and observe that the color of the light band changes.

6. If you change the wavelength far enough in either direction, you will not see any color because you will be beyond the region that your eye can detect. By careful observation, find out the limits for <u>your</u> eyes and record these on the data sheet.

7. Finally, find out which wavelengths correspond to which colors. To do this, return the wavelength to 400 nm. Look down the tube and record on the data sheet the color you see. Change the wavelength to 440 nm and record the color you see in the data table. Repeat this at 40 nm intervals from 400 to 720 nm. You should have recorded a color corresponding to each wavelength you selected.

Figure 5.2 Chalk prepared for the spectrophotometer

II. The Spectra of Red, Blue, and Green Solutions

1. Obtain at least two matched test tubes or other containers (sometimes called "sample cuvettes") that fit your spectrophotometer. Fill one of them about half full with a red dye solution. This will be the "sample." Fill the other tube half full with pure water. It will be the "blank." If you have a third tube available, fill it with a blue dye solution.

2. First adjust the "zero" on the spectrophotometer.

    If you have an analog Spectronic 20, with no tube in the sample holder and the *cover closed*, slowly turn the zero adjust knob (left front knob) until the meter needle or digital display shows exactly 0% transmittance (T). (For a meter with a needle, this requires placing your eye directly in front of the meter. If the meter has a mirror behind the needle, the mirror image of the needle should line up with the needle.)

    If you have a digital spectrophotometer, no adjust is required because the instrument automatically zeros itself.

3. Place the "blank" tube in the instrument and close the cover. Set the wavelength to 400 nm.

    If you have an analog Spectronic 20, carefully adjust the right-hand front knob (100% T Adjust) until the meter reads exactly 100% T or 0 absorbance (A). (Again, for best results, your eye should be directly in front of the meter.)

    If you have a digital spectrophotometer, press the "0 A" or "100 %T" button.

4. Now remove the "blank" tube, insert the "sample" tube with the red solution, and close the cover. *Without changing any knobs or pressing any buttons*, carefully observe the meter reading and record this in the data table as % T (or A) for the red solution at 400 nm.

5. If you have a third tube with a blue solution, place it in the instrument, close the cover, then read and record the % T (or absorbance) in the data table.

6. Now change the wavelength dial to 420 nm. Repeat steps 3 and 4 by first inserting the "blank," closing the cover, and adjusting the meter to read 100% T, or 0 A. Then insert the "sample," observe the meter reading, and record the % T or absorbance. (Repeat with the blue solution if possible.)

7. Proceed in the same fashion, at 20-nm intervals, all the way to 700 nm. This sounds tedious, but in fact, it goes quite rapidly, especially if two students work together, one of whom records the data while the other switches tubes. (Remember that <u>each</u> time you change the wavelength, you must insert the distilled water blank and adjust to 100% T, or 0 A. You then insert your sample and read the display without making any adjustments.)

8. Next, repeat the whole process with a blue solution if it was not done along with the red solution. Pour out the red solution, rinse the tube with water, and then fill it half way with a blue solution. Start at 400 nm and go up to 700 nm, recording the results in the data table.

9. You should stop now and make a graph of these results (see Part VI), or your instructor may have you simply examine the numbers to observe the pattern.

10. Finally, obtain the spectrum of a green solution. First, make a <u>prediction</u> of what you think the spectrum will look like for a green solution, based on your results with the red and blue solutions. Show it to your instructor before proceeding. Then rinse and fill the sample tube with a green solution and proceed to record the transmittance or absorbance data for this solution in the same manner as for the other solutions.

III. Optional Extension or Alternate Assignment: Study of An Acid-Base Indicator

As noted in the Background section, acid-base indicators are colored dyes that change color in acidic and basic solutions. It is interesting to measure the spectrum of such an indicator in both its acidic and basic forms—and even in its intermediate state halfway between the two forms.

If this assignment is done as an alternative to Part II, you can use the prepared table on the data sheet. If this is an additional study, you will need to construct a new table on a sheet of paper.

1. Rinse out the sample test tube and fill it halfway with the blue indicator solution provided (bromthymol blue). A small amount of sodium hydroxide has been added to this solution in advance to be sure the solution is alkaline. (If a third test tube is available, rinse and half-fill it with the same indicator solution. Then add a drop of hydrochloric acid solution to make the solution acidic. It should change color to yellow.)

2. Proceed in the same manner as in Part II. At each wavelength chosen, first insert the "blank" test tube and set the 100% T; then measure the colored solution. Start at 400 nm and go up to 700 nm. If you have test tubes with both forms of the indicator, measure both of them before proceeding to the next wavelength.

3. If only one sample test tube was available, add a drop of hydrochloric acid to the colored dye, mix by gentle swirling, and then make measurements from 400 to 700 nm.

4. **Challenge:** Find a way to make an intermediate form of the indicator solution that is green. You will have to experiment by adding small amounts of very dilute acid (hydrochloric acid) and very dilute base (sodium hydroxide) to achieve a green color. Predict what the spectrum

will look like, then measure the spectrum of this green solution and compare it with the spectra of the blue and yellow forms.

IV. Optional: Investigation of Colored Substances of Your Own Choosing

There are many colored substances to choose from. Some common examples include beverages (such as Kool Aid, some soft drinks, and juices), water solutions of jello, water-color paints, felt-tip pen inks, food color dyes, and hair dyes. You could even use a small piece of colored glass if it will fit in the test tube. As noted earlier, whatever you choose must be transparent (meaning that you can see through it), and it ought to have a fairly bright, distinct color.

You may have to experiment to find a suitable dilution of your substance so that the lowest transmittance does not drop below 5% T (or absorbance above 1.5). Then use the skills you have learned previously to collect data and plot a spectrum of the substance.

V. Optional: Ultraviolet or Infrared Spectra

Your instructor may demonstrate, or allow you to use, a spectrophotometer that measures ultraviolet or infrared spectra. You may be able to obtain an ultraviolet spectrum of a sunscreen material of the type used in Experiment 4. Or you may be able to obtain an infrared spectrum of one of the "greenhouse" gases such as carbon dioxide.

VI. Making Graphs of Your Results

Use the graph paper provided to plot the data for each colored solution or substance. The wavelength should be on the horizontal axis (x axis), ranging from 400 nm at the left edge to 700 nm at the right edge. Percent transmittance or absorbance should be on the vertical axis (y axis): % T should range from 0 at the bottom to 100 at the top. Absorbance should range from 0 at the bottom to 1.0 (or 2.0) at the top. First plot the red solution. *Using a pencil rather than a pen is preferable so that corrections can be made.* Carefully plot the data, making a small point with a circle around it for each measurement. When you are finished, draw a *smooth* curved line through the points as best you can. The line may not touch every point. (This is better than "connect the dots," which is likely to produce a somewhat jagged line.) Repeat this procedure with other colored substances. If you are careful, you can plot more than one spectrum on the same graph. Label each spectrum so that you know which is which. Your instructor will advise whether to plot the spectra separately.

Post-lab Questions

1. How does the energy content of light vary with wavelength (or color)? Consult Section 2.4 in the text for help. Which solution (red or blue) absorbed more light at higher energy wavelengths? Explain.

2. If you did the optional study of an acid-base indicator, which form (blue or yellow) absorbs light of highest energy? Explain briefly.

3. If you studied a sample of your own choosing, describe the sample and the spectrum. Is the spectrum approximately what you predicted?

Name _____   Date _____

Lab Partner _____   Lab Section _____

Data Sheet—Experiment 5

Wavelength limits for your eyes: lowest wavelength _____ highest wavelength _____

Wavelength (nm)	Color with chalk
400	

Wavelength nm	Red solution % T or Abs	Blue solution % T or Abs	Green solution % T or Abs	Student sample % T or Abs
400				

Notes

Name _____   Date _____

Lab Partner _____   Lab Section _____

Red solution

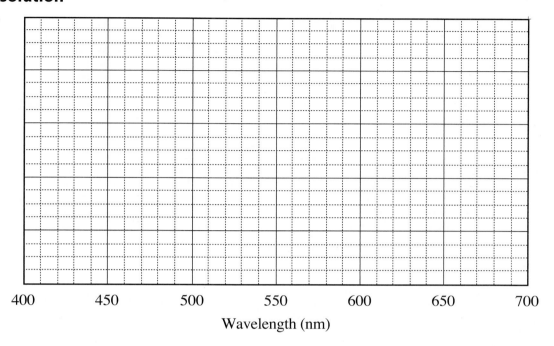

Wavelength (nm)

At what wavelengths does the solution have its greatest transmittance (lowest absorbance)? What color(s) of light correspond to these wavelengths?

At what wavelengths does the solution have its lowest transmittance (highest absorbance)? Here the dye molecules are absorbing the light. What color(s) of light is the red dye solution absorbing?

Blue solution

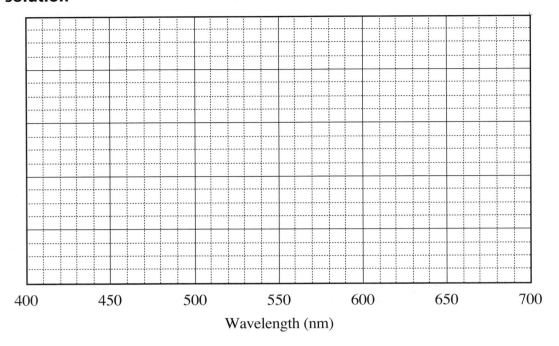

Wavelength (nm)

At what wavelengths does the solution have its greatest transmittance (lowest absorbance)? What color(s) of light correspond to these wavelengths?

At what wavelengths does the solution have its lowest transmittance (highest absorbance)? Here the dye molecules are absorbing the light. What color(s) of light is the red dye solution absorbing?

Propose a general rule relating the color of an object to the color(s) of the light that it absorbs.

Name _____   Date _____

Lab Partner _____   Lab Section _____

Green solution

Based on what you've done so far, predict what the spectrum of the green solution will look like. Show your answer to your instructor before proceeding.

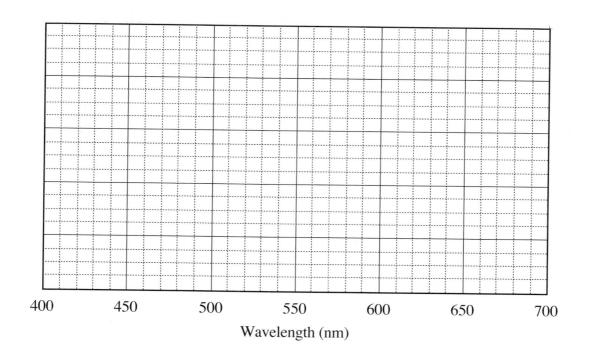

Did the spectrum of the green solution look as you predicted it would? Explain.

Additional graphs

You may use these graphs to record spectra from the acid-base indicator, to plot your dyes together, or for any other samples your instructor may assign.

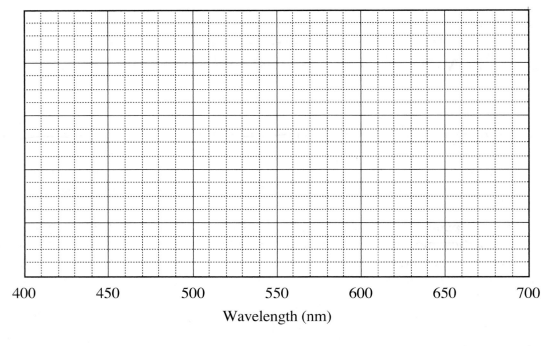

Wavelength (nm)

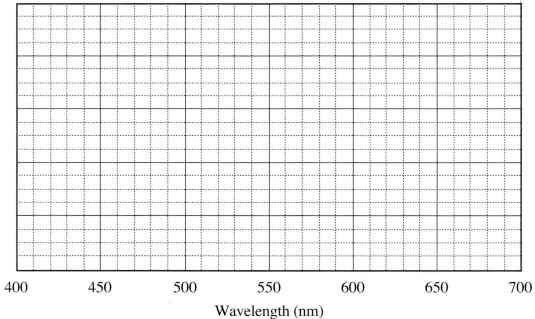

Wavelength (nm)

What Does a Molecule Look Like?
Bonds, Molecular Models, and Molecular Shapes

INTRODUCTION

The properties of chemical compounds are directly related to the ways in which atoms are bonded together into molecules. Section 2.3 of *Chemistry in Context* explains the connection between electrons and bonds, and how to write Lewis dot structures. Section 3.3 of *Chemistry in Context* shows how the three-dimensional shapes of molecules are related to the bonding. In this exercise, you will have the opportunity to apply your knowledge from those two chapters by constructing simple ball-and-stick models for some common molecules. The models should help your understanding of electron arrangements in molecules and the resulting shapes of the molecules. You will investigate a number of small molecules containing carbon, nitrogen, oxygen, and hydrogen, as well as a few molecules containing fluorine, chlorine, or sulfur. These are mostly substances that are important in the atmosphere and in polluted air, as discussed in Chapters 1, 2, and 3 of *Chemistry in Context*. In the process of doing this exercise, you will see how "models" become very useful to chemists in understanding and predicting chemical properties.

Background Information

The existence of chemical compounds with fixed composition implies that the atoms in compounds must be connected in characteristic patterns. Early models showed the atoms hooked together like links on a chain. Modern representations are a good deal more abstract and often mathematical in nature. Nevertheless, it is possible to represent molecular structures with reasonable accuracy by using relatively simple models. The models serve as a three-dimensional representation of an abstract idea. Molecular model building has proven so useful that it is rare to find a chemist who does not have a model kit close at hand.

The chemical bonds that hold atoms together in molecules generally consist of *pairs of electrons* shared between two atoms. Atoms tend to share outer electrons in such a way that each atom in the union (except hydrogen) has a share in an *octet of electrons* in its outermost shell. This generalization has come to be known as the **octet rule**. (You should review the discussion of Lewis structures and the octet rule in Section 2.3 of the text.) The location of each element in the periodic table provides information about the number of electrons in the outermost level of the atoms. Carbon, for example, is in Group 4A and has four outer electrons; thus, it must share four additional electrons from other atoms in order to achieve a share in eight outer electrons (an octet). Oxygen, in Group 6A, has six outer electrons and shares two electrons from other atoms in order to achieve an octet. Hydrogen is a special case, needing to share its one electron with only one electron from another atom in order to achieve the stable outer electron configuration of the nonreactive element helium (He). This is summarized in Table 6.1.

Table 6.1 Electron Configurations in Atoms and Molecules

Atom	Outer electrons	Electrons shared with another atom (bonds formed)
Carbon	4	4
Nitrogen	5	3
Oxygen	6	2
Fluorine & chlorine	7	1
Hydrogen	1	1

A **single bond** consists of one shared pair of electrons; a **double bond** is two shared pairs (i.e., 4 electrons), and a **triple bond** is three shared pairs (6 electrons). On paper, the bonds are represented by single, double, or triple lines, respectively (–, =, ≡). In most model kits, straight sticks represent single bonds, while double and triple bonds are represented by pairs or triplets of curved sticks or springs. Electrons not involved in bonding are termed *unshared electrons*.

An important part of this exercise involves identifying the *three-dimensional shapes* of molecules. (Molecular shapes are discussed in the text in Chapter 3.) Molecules have certain shapes depending on their component atoms and the ways in which they are bonded to each other. The important shapes encountered in this exercise are *linear, bent, triangular, pyramidal,* or *tetrahedral.* Several factors contribute to determining molecular shape. (1) Electron pairs (both shared and unshared) try to keep as far away from each other as possible, while still remaining "attached" to atoms. (After all, they are all negatively charged, and electrical charges of the same type will repel each other.) (2) Electron pairs tend to be symmetrically arranged around each atom in a three-dimensional manner. (3) Electron pairs *not* involved in the bonding ("unshared pairs" or "lone pairs") are equally as important as bonding electron pairs (shared pairs) in determining the overall molecular shape and arrangement of atoms.

Pre-lab Question
Draw Lewis structures for CH_4, NH_3, and H_2O. Which of these have lone pairs on some of the atoms? Can you predict the shape of each molecule?

EXPERIMENTAL PROCEDURE

Molecular model kits vary; therefore, your instructor will explain the particular models that you will use. The kit probably contains <u>balls</u> (used for atoms), <u>sticks</u> (used for single bonds and unshared electron pairs), and <u>springs</u> or <u>curved sticks</u> (used for double and triple bonds). Each stick or spring represents two electrons. Hydrogen atoms are usually represented by small, light-colored balls (yellow, white, or pale blue) that have only one hole. The color code for other atoms will vary. A common set of colors is shown in *Table 6.2*.

Note: There is one disadvantage to using the colored balls provided in most model sets. They usually have only enough holes for the correct number of bond pairs, and thus you will not be able to see the unshared electron pairs. An alternate approach is to use balls with four holes in them for all atoms other than hydrogen so that the octet (four pairs of electrons) will always be visible.

Table 6.2 Typical Color Code for Molecular Model Sets

Atom	Color
Hydrogen	Yellow or white
Carbon	Black
Nitrogen	Blue
Oxygen	Red
Fluorine or chlorine	Green
Bromine	Orange
Iodine	Purple
Sulfur	Yellow

Model Building Basics

1. Assemble the atoms required. (For example, to make the CH_4 molecule, you will need one carbon and four hydrogen atoms.) Next, note the group in the periodic table to which each element belongs. The number of the group is also the number of outer electrons in an atom of the element.

2. To determine how many sticks (pairs of electrons) you will need, divide the total number of outer electrons by 2. For example, H_2O has one outer electron from each hydrogen and six from oxygen, for a total of eight. Hence, you will need four sticks to represent all the electrons in H_2O. Two sticks represent bonds between H and O, and two sticks represent unshared electron pairs.

3. If there is only one atom of one element in the molecule and more than one atom of another element, the single atom usually goes in the center of the molecule. This is the case in CO_2, but there are a few exceptions to this rule (such as N_2O, which has the arrangement NNO).

4. With the collected parts, assemble the model in such a way that each atom except hydrogen has a share in an octet of electrons. If you do not appear to have enough sticks (electron pairs) to give each atom (except hydrogen) an octet, try sharing more electrons by forming double or triple bonds (replace straight sticks with curved sticks or springs).

The Assignment

1. Each pair of students should have a model set. First, get acquainted with the components of the set. Note the holes in the various colored balls and their positions. If there are sticks of several lengths or shapes, determine which are for single and which are for multiple bonds. If you have single bond sticks in two lengths, the short ones are for bonds involving hydrogen, and the longer ones are for any other single bond.

2. Using the procedure outlined above, build models for each of the molecules listed on the data sheet. Then use information obtained from viewing the models to fill in the information in the last two columns. You should take time to think about (and write down in words and a diagram) the shape of each molecule before proceeding to the next one.

Post-lab Questions

1. The tetrahedral shape is one of the most fundamental shapes in chemical compounds. How would you describe it in words to someone who has never seen it?

2. The octet rule appears to be a very important rule governing the structures of molecules. In light of your work with models, provide a simple explanation for the importance of eight electrons.

3. Explain in your own words how nonbonded electron pairs help determine the shapes of molecules.

4. CO_2 and SO_2 have very similar formulas. Did you find that they have the same geometry? Explain why, citing their numbers of total outer electrons. Did you find that SO_2 and O_3 have the same geometry? Explain why, citing their numbers of total outer electrons.

5. Do all of the assigned molecules obey the octet rule? If not, why (or in what way) did the octet rule fail? Can you suggest a general rule about the position of the atoms on the periodic table that do not obey the octet rule?

6. As a test of what you have learned, predict the shapes of (a) NF_3, (b) H_2S, (c) Cl_2O.

7. Models do not necessarily have to be physical objects. They can be two-dimensional drawings or even mental constructs. Cite one or more examples of such models encountered outside of chemistry. Can you think of models that are used in your own field of study or that you will use in your future career?

8. We often use computer-generated images to help us understand molecular structures. Consult your instructor to find out what molecular viewing programs are available at your institution. View some of the molecules that you studied in this experiment, and rotate them using the computer program. What advantages do you see for viewing molecules this way as compared to the pictures in your textbook? What advantages and disadvantages do you see for these computer images as compared to the physical models you constructed in lab?

9. You will revisit many of these molecules as you go through the course. Answer the following questions about some of the molecules you studied in this experiment.

    a. Define the term "greenhouse gas" and explain how CO_2 functions as one. What other molecules that you studied today can act as greenhouse gases?

    b. What role does O_3 play in the atmosphere? What molecules are involved in its formation?

    c. Explain the problems with the use of CFCs as refrigerants. What class of molecules has replaced them?

    d. What physiological role does NO play?

How Can We Measure the Mass of a Molecule?
Weighing Gases to Find Molar Masses

INTRODUCTION

Section 3.6 of *Chemistry in Context* introduces the concept of **molar mass**, which is defined as the mass (in grams) that contains one Avogadro number (6.02×10^{23}) of atoms or molecules of an element or compound. However, the text does not explain in much detail how molar masses are determined. This experiment provides an opportunity to actually measure molar masses of several gases and then to compare your experimental results with the accepted values for those gases.

Background Information

The principle behind this experiment is of fundamental importance in chemistry. The key idea is that equal volumes of gases, at the same temperature and pressure, contain equal numbers of molecules. Therefore, if we compare the masses of equal volumes of two gases, we will also be comparing the masses of equal numbers of molecules of the substances. The *ratio* of the masses of equal number of molecules will be the same as the ratio of the masses of *individual molecules* as well as also being the ratio of the *molar masses* of the substances involved. Stated mathematically, this can be written as

$$\frac{\text{mass of } n \text{ molecules of gas A}}{\text{mass of } n \text{ molecules of gas B}} = \frac{\text{mass of 1 molecule of gas A}}{\text{mass of 1 molecule of gas B}} = \frac{\text{mass of 1 mole of gas A}}{\text{mass of 1 mole of gas B}}$$

In this experiment, you will weigh equal volumes of several different gases. You will then correct these data to yield the actual masses of the gas samples. Finally, you will use the relationship above to relate the corrected masses to a standard reference substance (oxygen, O_2) and thus obtain values for the molar masses of the gases that you have investigated.

The experimental procedure that you will use consists of weighing equal volumes of several gases in a container made out of a plastic bag. Because the actual mass of the gas samples is at most a gram or two, great care must be taken in weighing the sample. Weighing a gas turns out to be more complicated than weighing a small solid object. All objects are buoyed up by the air that surrounds them, and if the volume of the air displaced weighs more than the object displacing it, the object floats. The most common example is a helium-filled balloon. If a helium balloon were placed on a balance, it would appear to weigh nothing! Even if an object does not float, it is still affected by an upward force equal to the mass of air it displaces or proportional to the volume of air it displaces. You, for example, appear to weigh about 100 grams (1/4 lb) less in air then you would weigh in a vacuum, depending on your volume. Of course your mass, the amount of matter you contain, is the same in air or in a vacuum. In order to get an accurate weight for an object, the buoyancy effect must be corrected for by adding the mass of the air displaced by the object to the weight of the object in air. For small solid objects, the correction is so small it can be ignored in all but the most meticulous measurements. However, when weighing gases, this buoyancy effect becomes significant. The way to correct your data for buoyancy will be explained in the calculation section of this experiment.

Overview of the Experiment

1. Prepare a plastic bag to use as a gas container.
2. Fill the container with several different gas samples, including one unknown gas.
3. Weigh the container filled with each of the gases.
4. Determine the volume of the container.
5. Determine the mass of a liter of air at room temperature and pressure.
6. Calculate the apparent and corrected mass of each gas.
7. Calculate the molecular mass and volume for each gas.
8. From the data, propose a possible formula for the unknown gas.

Pre-lab Question

Using values for atomic weights from the periodic table, calculate the molar mass of each gas you will measure in the experiment to the nearest whole number. Enter these values into the last column of the table on the data sheet.

EXPERIMENTAL PROCEDURE

I. Weighing The Gases

Caution! A single drop of water will add enough weight to significantly distort the results. Therefore, check carefully to make sure the bag is dry at the start and do all of the weighing of gases before you attempt to determine the volume of the bag.

Note about weighing: The most critical measurements in this experiment are weights (of the gas container plus gases), which are obtained by means of a **laboratory balance**. Your instructor will explain how to use the particular balance(s) available in your lab. It is important to remember, first, that the balance must always read exactly zero when nothing is on the balance pan. (It is a good idea to check it before each weighing.) Second, the balance readings should be recorded to at least the nearest 1/100th of a gram (0.01 g), i.e., the second digit to the right of the decimal point. (If the last digit is zero, be sure to record that.) Some balances may show a third decimal place; if so, include it. On the other hand, with some balances, it may be difficult to read the second decimal place with confidence, in which case you will have to make the best estimate you can.

1. Obtain a plastic bag, a one-hole rubber stopper with a medicine dropper stuck through the hole, a cork ring sized such that the rubber stopper fits snugly in the ring, and a dropper cap (see *Figure 7.1* on the next page).

2. Push the opened end of the plastic bag through the center of the cork ring; then open the bag and push the rubber stopper firmly into the open bag so that the bag is tightly caught between the stopper and the ring. Make sure there are no leaks in the system. To do this, inflate the bag, put the rubber cap over the hole in the medicine dropper, and gently squeeze the bag. No air should escape.

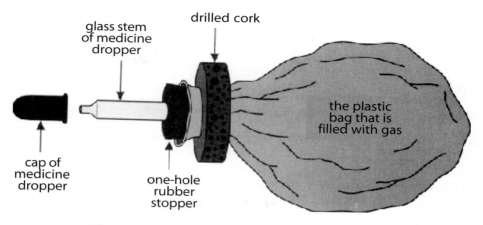

Figure 7.1 Diagram of the assembled system

3. With the rubber cap of the medicine dropper off, press out any air in the bag by smoothing it. It is important to get all the air out of the bag so that you weigh a completely empty bag.

4. Weigh the assembly, with the cap on the medicine dropper, to the nearest 0.01 gram and record the weight. (Remember to check first that the empty balance reads 0.00 g.)

5. Remove the cap from the medicine dropper and connect the bag via the medicine dropper to a source of gas.

6. Hold the bag by the stopper and completely fill the bag with one of the gases.

7. Allow the excess gas to escape so that the gas in the bag will be at room pressure but do not squeeze the bag. Replace the rubber cap.

8. Weigh the bag assembly containing the gas (which is at room temperature and normal atmospheric pressure) and record your result. Make sure that no part of the bag assembly hangs over the balance pan and touches any part of the balance frame or housing.

9. Completely empty the bag and repeat the procedure (steps 5–8) for the other gases available in the lab, including the unknown gas assigned to your team.

II. Finding the Volume of the Bag   (See *Figure 7.2*)

1. After you have weighed all the available gases, fill the bag with air at room pressure.

2. Attach a rubber hose to the bag at the medicine dropper.

3. Fill a large pan or the sink with water so that a large bottle of water can be inverted in it in such a way that the mouth of the bottle is below the surface.

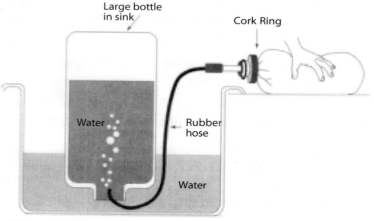

Figure 7.2 Diagram of the setup used to determine the volume of the plastic bag

4. Fill the large bottle with water (clear to the top, with no air space).

5. Cap the end of the bottle.

6. Invert the bottle so that the mouth is below the surface of the water in the sink or pan. (The bottle should be completely filled with water; no air bubbles should be inside.)

7. Remove the cap from the bottle while the bottle opening is underwater.

8. With the help of your lab partner, insert the free end of the rubber hose into the mouth of the bottle, as shown in the diagram.

9. Gently squeeze the air out of the plastic bag. As you do so, bubbles of air will enter the bottle and force out some of the water in the bottle.

10. When the bag is empty, keep the bottle inverted with the mouth under water while you replace and tighten the screw cap. Then remove the bottle from the sink or pan.

11. Place the bottle in a normal upright position and remove the cap.

12. Obtain a 1,000-mL (1-liter) graduated cylinder. Fill it with water to exactly the 1,000-mL line. Slowly pour water from the cylinder into the open bottle until the bottle is completely filled. Then record the volume of water left in the graduated cylinder. Subtract the volume of water remaining in the graduated cylinder from 1,000-mL. This is the volume of water that was required to completely fill the bottle. It is also equal to the volume of air that was originally in the bag.

13. Record the room temperature and air pressure. Your instructor may provide these values for the entire class, or you may be asked to measure them with a thermometer and a barometer.

14. Use Table 7.1 to determine the mass of 1 liter of air at the temperature and air pressure of the room. (**Note:** In the chart, air pressure is given in "mm Hg." This is the height of a column of liquid mercury in a mercury barometer. For comparison, standard atmospheric pressure at sea level is 760 mm Hg.)

III. Calculations (to be done for each gas sample)

1. Calculate the apparent mass of gas in the bag by subtracting the weight of the empty bag from the weight of the full bag. (This apparent mass may be less than zero. How can that be?)

2. Calculate the mass of air displaced by the bag of gas by multiplying the volume of the bag (in liters) by the mass of 1 liter of air at room temperature and pressure (taken from Table 7.1).

3. Calculate the corrected mass of the gas in the bag by adding the mass of the displaced air to the apparent mass of the gas.

4. Use the defined molar mass of oxygen (32 grams/mole) to calculate the experimental molar masses of the other gases, by means of the following equation.

$$\frac{\text{molar mass of gas A}}{\text{molar mass of oxygen B}} = \frac{x \text{ g/mol}}{32 \text{ g/mol}} = \frac{\text{measured mass of gas A}}{\text{measured mass of oxygen}}$$

5. Record your results in the table on the data sheet.

Post-lab Questions

1. One of the first ways that was used to determine the molar mass of a liquid compound utilized the following procedure. The liquid was placed in a glass container, and the container was heated until all of the liquid had boiled away. As a consequence, the air originally in the container was totally displaced by gaseous molecules of the unknown compound. The container was sealed (while still hot), then cooled and weighed. What information do you think was needed to complete the determination of the molar mass of the liquid?

2. Why can't we weigh a single molecule? Explain why finding the mass of a mole of those molecules is a good substitute.

Table 7.1 The Mass (in grams) of 1 Liter of Air
at Different Temperatures and Pressures

Pressure	Temperature			
(mm Hg)	15°C	20°C	25°C	30°C
600	0.97	0.95	0.94	0.92
610	0.99	0.97	0.96	0.94
620	1.00	0.98	0.97	0.95
630	1.02	1.00	0.99	0.97
640	1.03	1.01	1.00	0.98
650	1.05	1.03	1.02	1.00
660	1.07	1.05	1.03	1.01
670	1.08	1.06	1.05	1.03
680	1.10	1.08	1.06	1.04
690	1.11	1.09	1.08	1.06
700	1.13	1.11	1.09	1.07
710	1.14	1.12	1.11	1.09
720	1.16	1.14	1.12	1.10
730	1.18	1.16	1.14	1.12
740	1.19	1.17	1.15	1.13
750	1.21	1.19	1.17	1.15
760	1.22	1.20	1.18	1.16
770	1.24	1.22	1.20	1.18

Name _____   Date _____

Lab Partner _____   Lab Section _____

Data Sheet—Experiment 7

Room temperature (°C) _____   Atmospheric Pressure (mm Hg) _____

Weight of the empty bag assembly (in grams) _____

Final volume of water in the graduated cylinder _____

Volume of water added to the large bottle _____

Volume of the plastic bag when filled (in liters) _____

Weight of 1 liter of air at room temperature and pressure _____

Gas	Weight of bag assembly plus gas	Apparent mass of gas	Mass of air displaced	Corrected mass of gas	Calculated molar mass	Accepted molar mass
Oxygen					------	
Burner gas (methane)						
Carbon dioxide						
Argon						
Nitrogen						
Unknown						

The unknown gas contains some combination of carbon, hydrogen, fluorine, chlorine, or sulfur. Not all elements are present in any one unknown gas. Suggest one or more possible formulas for your unknown gas, assuming that the normal rules of chemical bonding for these elements are obeyed.

Compare your values for the molar mass of each gas with the accepted value. Do your results support the hypothesis that equal volumes of gases have equal numbers of molecules? Briefly explain the basis for your answer.

What are some likely sources of error that could account for the difference between measured and accepted values for molar mass? Indicate whether each one would make your answer too low or too high. (**Example:** How would the results of your experiment change if the temperature of one of the gas samples changed before it could be weighed?)

Chemical Moles: Baking Soda to Table Salt
How Do Chemical Equations Connect Compounds?

INTRODUCTION

During chemical reactions, substances combine with each other in a definite proportion by mass, meaning that only a certain amount of one reagent will react with a given amount of another reagent. The amounts of reactant species can be expressed in a variety of ways: grams, pounds, tons, and liters. However, no matter what units are used, they are all related to the ratio of *moles* of one species reacting with a certain number of *moles* of another species. If you are unclear about the concept and definition of a **chemical mole**, you should review the discussion of moles in Section 3.6 of *Chemistry in Context*.

In this experiment, you will investigate in more detail a reaction you have already used in Experiment 1, i.e., the reaction of sodium bicarbonate, $NaHCO_3$, with hydrochloric acid, HCl. Sodium bicarbonate is also known as sodium hydrogen carbonate, but it is most familiar as baking soda. The reaction of sodium bicarbonate with hydrochloric acid produces table salt (sodium chloride, NaCl), water, and carbon dioxide.

$$NaHCO_3 + HCl \rightarrow NaCl + H_2O + CO_2$$

In Experiment 1, you used this reaction to make carbon dioxide in a plastic bag. Now you will do quantitative measurements to find out how many moles of salt (NaCl) are formed from 1 mole of $NaHCO_3$. You cannot measure moles directly, but you will determine the mass of $NaHCO_3$ used and the mass of NaCl formed by weighing samples on a balance. To further improve the accuracy of the experiment, data will be assembled from the whole class, and you will be asked to decide on a "best" value for the ratio.

Overview of the Experiment
1. Label and weigh three test tubes.
2. Add a weighed quantity of sodium bicarbonate to each test tube.
3. React the sodium bicarbonate with 10% hydrochloric acid.
4. Evaporate the liquid remaining in the test tube after the reaction takes place.
5. Determine the weight of sodium chloride produced.
6. Calculate the ratio of moles of NaCl formed to moles of $NaHCO_3$ used.

Pre-lab Question
In the balanced equation for the reaction you will perform in lab, water (H_2O) and carbon dioxide (CO_2) are formed. These are not formed directly, but instead result from the decomposition of carbonic acid, which is formed first. What is the chemical formula for carbonic acid?

EXPERIMENTAL PROCEDURE

Note on the Use of a Laboratory Balance for Weighing: This experiment requires careful weighing of test tubes on a laboratory balance. Your instructor will explain how to use the particular balance(s) available in your lab. It is important to remember that the balance must always read exactly zero (0.00 g) when nothing is on the balance pan. It is a good idea to check it before each weighing.

I. Reaction

1. Obtain three test tubes that are clean and completely dry. Label them A, B, and C, <u>putting the labels near the tops of the tubes</u>. Add a small boiling chip to each test tube.

2. Take the test tubes plus the lab manual or data table to one of the laboratory balances. Weigh each of the test tubes (including the boiling chip) to the nearest hundredth of a gram (0.01 g) and record the masses in the data table, line 2. (If the second decimal place is a zero, be sure to record it, e.g., 18.10 g).

3. To each test tube add just enough $NaHCO_3$ to fill the curved bottom of the tube.

4. Again, weigh each test tube to the nearest hundredth of a gram (0.01 g) and record its mass (tube + contents) in the data table, line 1. **Note:** The masses of the three solid samples do not need to be identical. Typical masses of added $NaHCO_3$ will be about 0.3 to 0.7 g.

5. Fill a plastic pipet with 10% hydrochloric acid solution. Add this acid dropwise to tube A. Let the liquid run down the wall of the test tube and gently tap the tube after each drop reaches the bottom. Continue to add acid slowly until all of the solid has dissolved. (It is important to add only the <u>minimum</u> amount of acid needed to get the solid dissolved.) Save the tube and its contents for further work.

       CAUTION! 10% HCl is corrosive to the skin and other materials. Avoid spilling it on yourself, your partner, or your work space.

6. Repeat step 5 with each of the remaining test tubes (B and C).

7. <u>Gently</u> heat test-tube A and its contents over a Bunsen burner flame, holding the tube at an angle and pointed away from you (as well as pointed away from anyone else in the immediate vicinity). See *Figure 8.1*. The idea is to evaporate the water in the tube without spattering anything out of the tube.

       STOP! No flammable chemicals should be in the vicinity of the Bunsen burner. Long hair should be tied back and extremely loose sleeves on clothing should be avoided.

    NOTE: Too rapid heating of the tube, especially if it is held in an upright position, will cause the hot contents to splash out of the tube and will necessitate starting over with a fresh sample. Boiling chips should help to produce smooth boiling. Your instructor may provide additional advice on how to minimize the problem.

    Continue heating until *all* of the liquid has evaporated and solid NaCl remains. (It is crucial to the success of this experiment to be sure that all of the water has evaporated from the *upper* part of the tube.)

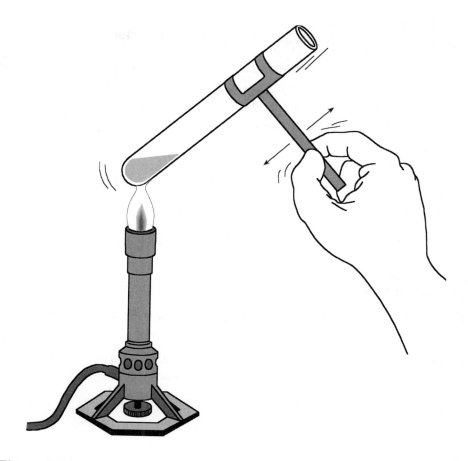

Figure 8.1 The correct way to heat a test tube over a Bunsen burner

8. Remove the tube from the flame and test for the evolution of water vapor from tube A by inverting a clean, dry test tube over the upright mouth of test-tube A. If condensation occurs in the cold test tube, continue the drying and testing process until no condensation occurs. Then set test-tube A with its dried contents aside to cool.

9. Repeat the procedures in steps 7 and 8 with test-tubes B and C.

10. Allow the three test tubes to cool (at least 5 minutes), *check to be sure there are no water droplets left*, and then weigh each with its contents. Record the masses in the data table.

11. If time permits, confirm that the tubes were fully dried by reheating them for 1–2 minutes, cooling for 5 minutes and reweighing. Record the second mass for each. If they were dry the first time, there should be a negligible change in mass.

Clean-up

Dispose of the test tube contents as your instructor directs you. Do not put anything down the sink drain without being told that is permitted. Thoroughly wash the test tubes, rinse with deionized water, and leave them upside down to drain.

II. Optional Extension or Alternate Assignment

After completing this study with sodium bicarbonate, you might wish to try a similar study with sodium carbonate (Na_2CO_3) to see how the results differ. Your instructor may ask some or all members of the class to do the experiment with sodium carbonate instead of sodium bicarbonate. The experimental procedure is the same, and the calculations are almost the same. If you do this version, you will need to make the appropriate changes in lines 1, 3, 4 and 9 of the data table.

III. Calculations

1. For the three columns in the data table, subtract the numbers on line 2 from line 1 to obtain the masses of $NaHCO_3$ used (line 3).

2. Add up the molar masses of the atoms in $NaHCO_3$ to find the mass of 1 mole of $NaHCO_3$. Then use this result and the masses on line 3 to calculate the number of moles of $NaHCO_3$ used (line 4).

3. Line 6 (mass of tube) is the same as line 2. Subtract line 6 from line 5 to obtain the masses of $NaCl$ formed by the reaction (line 7).

4. Proceeding the same way as in step 2, calculate the moles of $NaCl$ formed (line 8).

5. Finally, calculate the <u>ratio</u> of moles $NaCl$ to moles $NaHCO_3$ (line 9). This should be recorded to two decimal places (i.e., 2 digits after the decimal point).

6. Calculate the average of your three experimental results for this ratio and record it to two decimal places.

Post-lab Questions

Balance each equation in the mechanism showing formation of ozone (O_3) in the upper atmosphere.

a. First, nitrogen and oxygen gas react to form nitrogen oxide.

$$N_2 + O_2 \rightarrow NO$$

b. Then, nitrogen oxide reacts with more oxygen to form nitrogen dioxide.

$$NO + O_2 \rightarrow NO_2$$

c. When nitrogen dioxide is struck by UV light, it decomposes to form nitrogen oxide and a highly reactive oxygen atom.

$$NO_2 + UV \text{ light} \rightarrow NO + O$$

d. Finally, the oxygen atom reacts with diatomic oxygen to form ozone.

$$O + O_2 \rightarrow O_3$$

e. How many moles of O_2 are required to form one mole of O_3?

Hot Stuff: An Energy Conservation Problem
Can I Measure What I Can't See?

INTRODUCTION

This laboratory exercise is a departure from the usual experiment. It is an attempt to simulate the kind of problem solving that takes place in a scientific laboratory. There are no instructions, procedures, or data sheets. There is simply a problem to solve that requires reasoning skills and the application of previous experience or knowledge.

A student team will typically consist of three or four members, one of whom should be designated as the recorder for the group. This person records ideas, data, experimental procedures, etc. Your group might also want to assign other tasks to members of the group, for example, a "go-fer," an experimentalist, a report writer, a team leader. (Your investigation will be more efficient if you take a few minutes at the beginning to be sure each person has a clear role to play.) Your team must devise a method for solving the problem and then obtain a numerical answer. When finished, your instructor will lead a class discussion in which teams will compare answers and methods.

THE PROBLEM

As an energy conservation method, you decide to turn down your home water heater so that it only heats water to 55°C. (It previously heated the water to between 60°C and 70°C.) Unfortunately, when you decide to measure the temperature of the water in your water heater, the only thermometer available has a maximum temperature of 40°C. In the laboratory, there is a simulated water heater, consisting of a large coffee pot filled with hot water. Each team will have available a thermometer that has the graduations above 40°C covered so that you cannot see them. Using your 40°C thermometer and materials available to you in the lab, devise a way to measure the temperature of the water in the coffee pot. Your team should be prepared to defend your answer and the method you used to obtain it.

Optional: Your instructor may ask you to propose two or three *different* methods for solving this problem, then test each of them, decide which gives the most reliable answer, and explain why.

Materials and Equipment

A variety of materials and equipment will be available that you can use to solve this problem. At a minimum your "lab" should have graduated cylinders, burets, beakers, flasks, plasticware, Styrofoam cups, stirrers, test tubes, plastic pipets, a balance, and, of course, a 40°C thermometer.

Reporting Your Results

When everyone is finished, your class will assemble to hear a report from each student team about the method used and the results. Therefore, it is important for your team to keep a complete record of everything you do and the numerical data you obtain. In addition to an oral report to the class, your group should prepare a brief (one-page) written report of your investigation. Your instructor will specify what should be included in the report.

Notes

Which Fuels Provide the Most Heat?
Comparison of the Energy Content of Fuels

INTRODUCTION

The ready availability of high-quality energy sources is one of the most important issues facing the global community in the late twentieth century. Chapter 4 in *Chemistry in Context* examines some of the societal choices we face and the advantages/disadvantages of various fossil fuels. Hydrocarbons (and mixtures of hydrocarbons) are among the most energy-rich fuels. Many of them are liquids, which makes them attractive for transportation fuels.

In this experiment, you will investigate the energy content of several fuels by using them to heat water. The data that you and your classmates obtain will enable you to compare fuels to see which ones provide more energy for a given mass of fuel burned. In particular, you will be able to determine how the energy released by burning hydrocarbons compares to that released by burning oxygenated fuels such as ethanol, a topic discussed in Section 4.9 of *Chemistry in Context*.

Background Information

Fuels are substances that burn to give off relatively large amounts of heat. In an overall sense, such burning is simply a combustion reaction between the fuel and oxygen. The amount of heat generated depends on what kind of fuel is used and how much of it is burned.

The simplest example of a common fuel is natural gas, which is almost pure methane (CH_4). When methane is burned completely, the only products are carbon dioxide and water.

$$CH_4 + 2\,O_2 \longrightarrow CO_2 + 2\,H_2O$$

Methane is an example of a hydrocarbon fuel. **Hydrocarbons** are compounds that contain only hydrogen and carbon. Other hydrocarbons, such as propane (C_3H_8), which is used in barbecue grills, and butane (C_4H_{10}), used in welding torches, will burn to produce the same products, carbon dioxide and water.

Some fuels such as ethanol (ethyl alcohol, C_2H_5OH) contain oxygen in addition to carbon and hydrogen. In effect, they are hydrocarbons that have already partially reacted with oxygen:

$$2\,C_2H_6 + O_2 \longrightarrow 2\,C_2H_5OH$$

In this experiment, you will measure the amount of heat given off by known amounts of several fuels. Your instructor will tell you exactly which fuels will be studied and how the class assignments will be made. Some of the fuels will be alcohols such as methanol (CH_3OH), ethanol (C_2H_5OH), isopropanol (C_3H_7OH), or butanol (C_4H_9OH). Some will be hydrocarbons, such as lamp oil or candle wax. Each of the latter is actually a mixture of hydrocarbons, but we will approximate their compositions as $C_{12}H_{26}$ (lamp oil) and $C_{40}H_{82}$ (candle wax).

The experimental procedure is to heat some water by burning a measured amount of a fuel sample. It takes exactly 1 calorie (cal) of heat to raise the temperature of 1 gram of liquid water by 1°C. (In Section 5.2 of the text, this is called the "specific heat.") Therefore, if you know the mass of water and how many degrees the temperature goes up, then the total amount of heat absorbed by the water can be calculated as follows:

$$\text{heat absorbed (calories)} = \text{mass of water (grams)} + \text{temp. change (C)} + 1.00 \text{ cal/g°C}$$
$$= m + \Delta T + 1.00 \text{ cal/g°C}$$

ΔT is a shorthand way of saying "change of temperature." In this equation, the last term (1.00 cal/g°C) is the specific heat of water and is used to make the units come out right. (You should convince yourself that combining and canceling the units on the right side will leave only calories.) Theoretically, the amount of heat liberated by the burning fuel should equal the heat absorbed by the water, but in practice, some of the heat will be lost to the surroundings.

Overview of the Experiment

1. Assemble the apparatus and obtain a burner containing a known fuel.
2. Add a measured volume of water to the can and then determine the mass of the water.
3. Weigh the burner.
4. Record the initial temperature of the water.
5. Light the burner and heat the water until the temperature increases about 20°C.
6. Record the highest temperature of the water.
7. Weigh the burner again, in order to find the mass of fuel used.
8. Repeat with two or more additional trials.
9. For each trial, calculate the amount of heat released per gram of fuel burned.

Pre-lab Question

Which fuel do you think would release the most energy upon burning, ethane (C_2H_6) or ethanol (C_2H_6O)? Why?

EXPERIMENTAL PROCEDURE

The following general procedure is to be used for each measurement. Your instructor will tell you which fuel or fuels you are to investigate. You may be asked to do one trial with each of three different fuels. Alternatively, you may be asked to do several trials with a single fuel to improve the level of your confidence about an average value for that particular fuel, in which case the results for the whole class will be assembled for comparison.

Important Note About Use of Laboratory Balances: The success of this experiment depends heavily on the accuracy of many mass measurements. All weighing will be done on a laboratory balance, which you will be sharing with other students. Your instructor will explain the use of the particular balances in your lab. For each weighing, it is important to know that the balance reads *exactly zero* with nothing on the balance pan. For some kinds of laboratory balances, the "zero" can be changed easily, and perhaps unintentionally, so your measurements might be in error. It is important to check the "zero" *each* time you make a weighing.

I. Testing the Fuels

 STOP! You need to be aware constantly that you (and other students) are working with flammable solvents and open flames. Handle the fuel burners very carefully. Before starting the experiment, be sure you know where a fire extinguisher is located in your laboratory. As with any experiment involving open flames, long hair should be tied back and extremely loose sleeves should be avoided. Wear eye protection at all times.

1. Obtain a dry soda can with the top removed and two holes punched on opposite sides near the top. Slide a glass rod through the holes in the can so that it can be suspended from the ring attached to a ring stand as shown in the diagram on the next page.

2. Obtain a fuel burner, place it under the can, and adjust the height of the ring so that the bottom of the can is about 2 centimeters above the top of the wick. (*Figure 10.1.*)

3. Take the empty soda can plus your data sheet to a balance. Check to be sure that the empty balance reads 0.0 gram; then weigh the can and record the mass to the nearest 0.1 gram.

4. Using a graduated cylinder, add approximately 100 mL of water to the can. Reweigh the can plus water to the nearest 0.1 gram. By subtraction, calculate the mass of water in the can.

5. Put a thermometer in the can, stir the water for a few moments, and then measure the temperature of the water, trying to estimate to the nearest 0.1 degree Celsius. (If you are using a glass thermometer, it is probably marked at 1 degree intervals, so it is not possible to measure the temperature accurately to a tenth of a degree. Nevertheless, it is useful to make the best estimate that you can.)

6. Take the fuel burner plus your data sheet to a balance. If necessary, check to be sure that the empty balance reads zero. Weigh the burner and record its mass on the data sheet, reporting it to the nearest 0.01 gram (or the nearest 0.001 gram if the balance permits this).

    Note that the can and water need only be weighed to the nearest 0.1 gram, while the burner must be weighed to the nearest 0.01 or 0.001 gram.

7. Place the fuel burner under the soda can and light the burner. Observe the flame. If necessary, cautiously adjust the height of the can so that the top of the flame is just below the bottom of the can.

8. Stir the water occasionally and continue heating the water until the temperature has increased about 20 degrees Celsius; then extinguish the flame.

Quickly do the following two steps:

9. Continue stirring the water gently until the temperature stops rising; then record the highest temperature, again estimating it to the nearest 0.1 degree. Calculate the temperature change by subtracting the initial temperature from the final (highest) temperature.

10. Take the burner and your data sheet to a balance. If necessary, check that the balance reads zero and then weigh the burner and record the mass to the nearest 0.01 gram (or 0.001 gram). Calculate the mass of fuel burned by subtracting the final weight of the burner from the original weight of the burner.

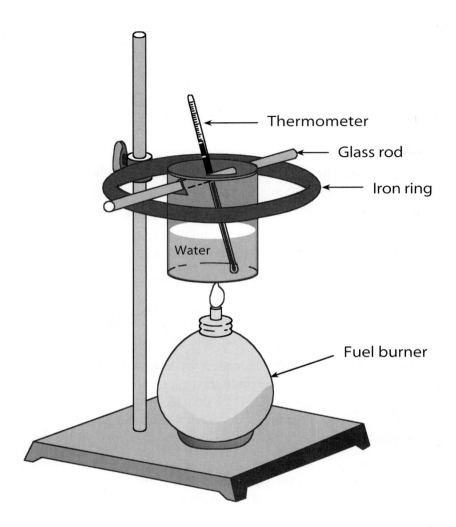

Figure 10.1 Diagram of can and burner setup

11. You should then do the following calculations, using the measurements you have recorded on your data sheet and the equation given in the introduction:

   a. For each trial, calculate the total calories of heat absorbed by the water. This will be assumed to be equal to the amount of heat liberated by the burning fuel.

   b. Then calculate the calories of heat per 1 gram of fuel burned.

   c. **Optional:** You may be asked to calibrate your measurements using methanol

   Do these calculations for each trial as soon as it is finished to see how the results are coming out.

12. Before doing another measurement, take a few moments to discuss the procedure with your partner. Did you encounter any difficulties? Can you think of any desirable improvements in the procedure? Also do the calculations, as described in the next section. This will give you a better idea of what you are looking for in subsequent trials.

13. Now repeat the procedure, either making additional measurements with the same fuel or switching to different fuels, as directed. If the water in the can has cooled to nearly room temperature, you can use the same water sample. If not, it is a good idea to empty the can and add a fresh 100-mL portion of water. Note that you recorded the mass of the *dry* empty can before you started. This will not change, except for possible buildup of soot on the bottom; therefore, you need to measure only the mass of the can with water in it.

II. Interpretation of Results

Because it was only possible for you to do a few trials, it is desirable to assemble a much larger body of data from your whole class or lab section so that more reliable comparisons of fuels can be made. Your instructor will tell you how to post or report your results for the rest of the class to see. You may be asked to calculate class averages for each fuel that was used. The following questions should help to focus your interpretation of the results.

III. Optional Extensions

1. **Investigate the effects of changes in procedure.** These might include (a) adding some nonflammable insulation around the can, (b) adding a cylindrical shield of aluminum foil around the burner, (c) using 200 mL of water and only a 10°C temperature rise (or 50 mL of water and a 40°C temperature change). Predict what will happen with each change and evaluate your hypothesis based upon your observations.

2. **Energy content of wood.** See Performance-Based Assessment Activity #3.

3. **Caloric content of nuts.** See Performance-Based Assessment Activity #4.

Post-lab Questions

1. Gasohol is a mixture of gasoline (hydrocarbons) and ethanol (an alcohol).
   a. How would the energy content per gram of gasohol compare to that of plain gasoline?
   b. How do you think this would affect a car's fuel efficiency (miles per gallon)?
   c. Consider that gasoline comes from petroleum, while ethanol can be made from corn.
      - Explain why environmentalists might promote the use of gasohol.
      - Explain why farmers and farm advocates might promote the use of gasohol.
      - Explain how someone concerned with feeding the poor might feel about increasing use of gasohol.
   d. Because of concerns about the effects of producing fuel ethanol from corn on global food prices, there is currently a lot of research into alternative raw materials for ethanol production. Go online and search for information about other natural resources that can be used for making ethanol. List a few of these materials and explain some of their advantages. Be sure to cite your sources.

2. Carbohydrates have the general formula CH_2O, while fats typically have a general formula of about $C_{10}H_{19}O$. Which of these classes of foods has the greater oxygen content as a percentage of the molecule's mass? Since foods are simply fuels for the body, which of these will release more energy when metabolized in the body? Explain your reasoning. How does this conclusion compare to the relative calorie content of fats and carbohydrates? (If you are not sure how many dietary Calories are in a gram of fat or carbohydrate, see Section 11.6 in your textbook.)

3. Suppose you put 50 mL of water in the can instead of 100 mL and heated it 40 degrees instead of 20 degrees. In what ways, if any, would this affect the results of the experiment? What would happen if you used 200 mL of water and heated it only 10 degreees? Can you think of any advantages or disadvantages in using either 50 mL or 200 mL of water in this experiment?

4. There are many possible sources of error in this experiment. List three that you can think of. Would each error have a large effect, a medium effect, or a small effect on the calculated heat content of a fuel? Also indicate whether the calculated result would be too high, too low, or could go either way.

Can Waste Oil be Turned into a Fuel?
Biodiesel: Preparation and Properties

INTRODUCTION

Because our stockpiles of fossil fuels are dwindling rapidly, there is active research into alternative and renewable fuels. As discussed in Chapter 4 of *Chemistry in Context*, biofuels are derived from natural sources such as plants and trees. Indeed, ethanol and biodiesel are two types of biofuels that have become commonplace. While each certainly has environmental benefits compared to petroleum, environmental and social justice concerns have been raised regarding these fuels and much research is being directed toward reducing the environmental footprint of biofuels, including producing them from waste rather than new materials.

Background Information

Turning waste materials into useful products offers both economic and ecological benefits. Waste oils from cooking can be burned directly for heat, but they don't have the right characteristics to substitute for gasoline or diesel fuel in cars. Fortunately, a simple chemical reaction will convert oils to a more useful form called biodiesel. Either waste or new fats and oils derived from plants or animals can be utilized for biodiesel production. Used oil contains waste products that complicate the procedures by requiring the oil to be treated prior to biodiesel production, so you may use new oil or treated waste oil for this experiment at the discretion of your instructor.

As discussed in section 11.3 of the textbook, oils and fats are composed of triglycerides like the one shown in the equation below. Triglycerides are even larger molecules than they appear in the equation because the three R-groups shown in the triglyceride each represent a long hydrocarbon chain with a length of 10-20 carbon atoms with hydrogens attached.

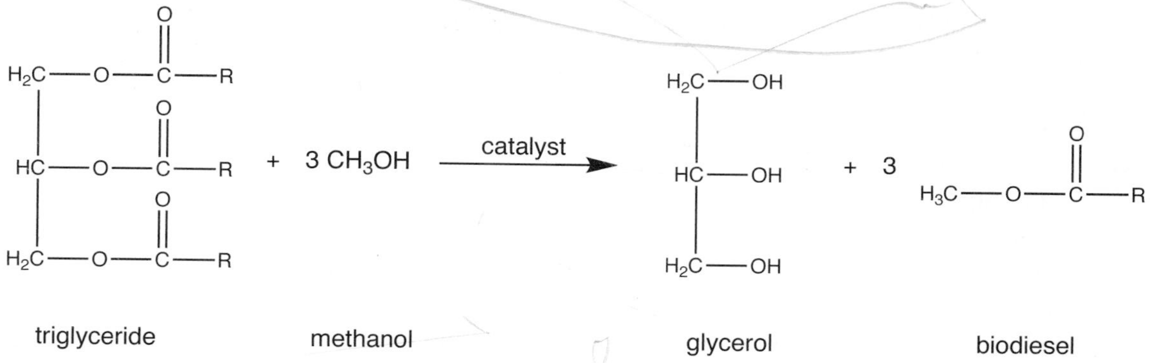

triglyceride   methanol   glycerol   biodiesel

Figure 11.1 Synthesis of Biodiesel

As depicted in *Figure 11.1*, the triglyceride molecules react with methanol in the presence of a catalyst (NaOH) to form two products. The first is glycerol, which is useful for making cosmetics and other consumer products. However, due to biodiesel production there is currently a glut of glycerol and research is being directed into something useful to do with it all, including turning it into biodegradable plastics. The second is biodiesel, a methyl ester (discussed in section 10.3 of the textbook) of one of the long chain fatty acids that was removed from the triglyceride. Vegetable oils contain mixtures of several triglycerides, where the number of carbon atoms in the R-group varies. As a result, any biodiesel fuel will also be a mixture of several molecules each with a different number of carbon atoms in R. Each type of vegetable oil will have a different mixture of triglycerides and give us a different mixture of biodiesel molecules. This means that biodiesel fuels prepared from different types of oils will have somewhat different properties. In this experiment, you will examine several characteristics of the biodiesel you prepare that are important to its use as a fuel, including viscosity, gelation temperature, and energy content.

Viscosity is a measure of how easily a liquid flows or pours. For instance, water has a low viscosity and flows easily, while maple syrup has a much higher viscosity and pours slowly. Because a fuel must flow through the engine, its viscosity is important to the proper functioning of that engine. If the fuel's viscosity is too high, the engine will have to work harder, causing higher operating temperatures and greater wear. You will measure the viscosity of both your biodiesel product and the oil from which it was prepared by measuring how long it takes each liquid to drip from a small pipet. The longer it takes for the liquid to drain, the higher the viscosity of the liquid.

Diesel engines are notoriously difficult to start in cold winter weather. The problem is that the diesel fuel thickens and turns to a gel at these temperatures. Gelled fuel will not flow or ignite properly. You will observe the biodiesel and oil to see if they begin to cloud and gel at typical winter temperatures, near 0°C.

Of course, the most important characteristic of a fuel is that it burns to release energy. If you did Experiment 10, you have already measured the energy content of several fuels. In this experiment, you will follow that same procedure to measure the energy content of your biodiesel. You will burn some of your biodiesel and use the heat released to warm a known amount of water. The temperature increase of the water will allow you to calculate the number of calories of heat released. Dividing this by the number of grams of fuel burned will give you a measure of the available energy of the fuel in calories of heat per gram of fuel burned. Consult Experiment 10 for more on related concepts.

Overview of the Experiment
1. Mix methanol and NaOH
1. Heat oil to 50°C.
1. Add methanol mixture and stir for 20 minutes.
1. Centrifuge product and then remove top biodiesel layer for testing.
5. Measure the viscosity of the oil and the biodiesel.
6. Observe the effect of low temperatures on the oil and the biodiesel.
7. Add a wick and some biodiesel to a 20 mL beaker.
8. Follow the procedure of Experiment 10 to measure the energy content of your biodiesel.

Pre-lab Question

How does biodiesel compare to petrodiesel in origin and chemical composition?

EXPERIMENTAL PROCEDURE

I. Preparation of Biodiesel

1.   If your class is preparing biodiesel from several types of oil, your instructor will tell you which to use. Record the type of oil.

2.   Use a 10 mL graduated cylinder to measure out 10 mL of methanol into a small beaker.

3.   Add 0.5 mL (about 10 drops) of 9M NaOH to the methanol. Record observations. Set this mixture aside until step 7.

   CAUTION! NaOH is caustic and can harm skin. If you spill any on your hands, wash them immediately and notify your instructor.

4.   Use a 50 or 100 mL graduated cylinder to measure out 50 mL of your oil. Pour the oil into a 250 mL beaker.

5.   Add a magnetic stirring bar to your oil and place the beaker on a stirring hotplate. Adjust the stirring rate so that the solution is being well mixed without splashing. Turn the heat on <u>low</u>.

6.   Heat your mixture to 50°C. To test the mixture's temperature, hold a thermometer in the mixture but well away from the spinning stir bar. Do not stand the thermometer in the beaker because it will break if struck by the stir bar.

7.   As soon as your oil reaches 50°C, <u>turn off the heat</u>, but keep the oil stirring. Slowly pour your methanol/NaOH mixture into the oil. Be sure that the mixture continues to stir sufficiently so that the methanol does not form a layer on top. Record observations.

8.   After 20 minutes, remove your beaker from the hotplate. Record observations.

9.   Obtain four centrifuge tubes. Fill each three-quarters full with your biodiesel mixture. Place the tubes opposite of each other in a centrifuge and spin them for several minutes.

10.   Turn off the centrifuge and allow it to stop before trying to remove your tubes. You should see two well-separated layers. If not, return the tubes to the centrifuge for several more minutes. The top layer is your biodiesel. The bottom layer is glycerol.

11.   Use a disposable pipet to transfer the top biodiesel layers to a small beaker. You will need about 20 mL of biodiesel in total. If you don't have this much, pour more of the reaction mixture from the 250 mL beaker into centrifuge tubes and repeat the separation process.

12.   Keep the small beaker of biodiesel for testing. Discard the lower layer of glycerol and any of the mixture remaining in the 250 mL beaker as directed by your instructor.

II. Viscosity of Oil and Biodiesel

1.   Obtain two 6-inch disposable glass pipets that have been marked with lines on the barrels 4 cm apart, as shown in *Figure 11.2*. You'll also need a matching dropper bulb and a timer that can measure in seconds.

2. We will estimate viscosity by measuring the time required for the liquid level in the pipet to drop from the upper line to the lower line. Viscosity is proportional to this time.

3. Use the bulb to fill one of your pipets with your biodiesel so that the liquid is above the upper line. Hold the pipet vertically over the beaker of biodiesel. Quickly remove the bulb and place your index finger firmly over the top of the pipet. This will stop the biodiesel from dripping.

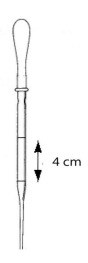

4 cm

4. Have your partner ready to time. Remove your finger from the top of the pipet. Allow the liquid to drain freely back into the beaker. Measure the time required for the biodiesel level to drop from the upper line to the lower line. Record your time.

5. Repeat the measurement. Record your data and calculate the average. Discard the used pipet.

Figure 11.2
Pipet for viscosity measurement

6. Use a test tube to get about 5 mL of the oil used in Part I as the starting material for your biodiesel. Use your second marked pipet to measure the viscosity of the oil. Allow the oil to drip back into its test tube. Do the measurement twice, recording your data and calculating the average time. Save your oil for Part III.

III. Temperature Effect on Oil and Biodiesel

1. Fill a 250 mL beaker most of the way with ice. Add two large scoops (at least 2 teaspoons full) of salt to the ice and then add enough water to make the mixture slushy. Stir briefly to mix in the salt. The presence of salt lowers the temperature of an ice bath

2. Place a thermometer inside your test tube of oil. Note how cloudy or clear the oil appears. Note how viscous (thick) the liquid appears as it drips from the thermometer or flows down the inside wall of the test tube. Now place the test tube of oil in your ice bath.

3. Observe the oil as it cools. Periodically remove it from the ice long enough to see if the oil is becoming cloudy or more viscous. If so, record the oil's temperature. Keep the test tube in the ice until its temperature is no longer dropping. Again note the oil's appearance and apparent viscosity. Record your observations.

4. Wipe the oil from your thermometer with a paper towel or tissue.

5. Now put some of your biodiesel in a test tube, add the thermometer, and repeat the cooling and observation process.

Cleanup

Discard or recycle the contents of your test tubes as directed by your instructor. Clean your thermometer with soap and water to remove all traces of the oil.

IV. Heat Content of Biodiesel

1. Prepare a burner for your biodiesel. Obtain a 20 mL beaker, a candle wick, and a paper clip. The end of the candle wick should be wrapped around one wire of the paper clip. Spread the paper clip out slightly so that it makes a stable bottom support for the wick. The wire inside the wick will allow it to stand vertically. Put the paper clip and wick in your

beaker. The wick should extend to the top of the beaker. Now fill the beaker most of the way with your biodiesel fuel.

2. The wick should stand freely in the middle of the biodiesel. Use matches to light the wick and ensure that your burner is working well. Once you know that your burner is working, extinguish its flame.

3. Obtain a dry soda can with the top removed and two holes punched on opposite sides near the top. Slide a glass rod through the holes in the can so that it can be suspended from the ring attached to a ring stand as shown in *Figure 10.1* in Experiment 10.

4. Take the empty soda can plus your data sheet to a balance. Check to be sure that the empty balance reads 0.0 gram; then weigh the can and record the mass to the nearest 0.1 gram.

5. Using a graduated cylinder, add approximately 100 mL of water to the can. Reweigh the can plus water to the nearest 0.1 gram. By subtraction, calculate the mass of water in the can.

6. Put a thermometer in the can, stir the water for a few moments, and then measure the temperature of the water, trying to estimate to the nearest 0.1 degree Celsius. (If you are using a glass thermometer, it is probably marked at 1 degree intervals, so it is not possible to measure the temperature accurately to a tenth of a degree. Nevertheless, it is useful to make the best estimate that you can.)

7. Take the biodiesel burner plus your data sheet to a balance. If necessary, check to be sure that the empty balance reads zero. Weigh the burner and record its mass on the data sheet, reporting it to the nearest 0.01 gram (or the nearest 0.001 gram if the balance permits this).

   Note that the can and water need only be weighed to the nearest 0.1 gram, while the burner must be weighed to the nearest 0.01 or 0.001 gram.

8. Place the biodiesel burner under the soda can and light the burner. Observe the flame. If necessary, cautiously adjust the height of the can so that the top of the flame is just below the bottom of the can.

9. Stir the water occasionally and continue heating the water until the temperature has increased about 20 degrees Celsius; then extinguish the flame.

10. Continue stirring the water gently until the temperature stops rising; then record the highest temperature, again estimating it to the nearest 0.1 degree. Calculate the temperature change by subtracting the initial temperature from the final (highest) temperature.

11. Your burner is probably hot to the touch near its top, so handle the beaker at its base. Take the burner and your data sheet to a balance. If necessary, check that the balance reads zero and then weigh the burner and record the mass to the nearest 0.01 gram (or 0.001 gram). Calculate the mass of fuel burned by subtracting the final weight of the burner from the original weight of the burner.

12. Now repeat the procedure. Discard the warm water from your can and repeat steps 5-11. Note that you recorded the mass of the *dry* empty can before you started. This will not change, except for possible buildup of soot on the bottom; therefore, you need to measure only the mass of the can with water in it.

Calculations

Calculate the heat released by your biodiesel for each trial as directed in Experiment 10, then average the results of your trials together.

Interpretation of Results

Because it was only possible for you to do a few trials, it is desirable to assemble a much larger body of data from your whole class or lab section so that more reliable comparisons of biodiesel fuels can be made. Your instructor will tell you how to post or report your results for the rest of the class to see. You may be asked to calculate class averages for each biodiesel if different oils were used. The following questions should help to focus your interpretation of the results.

Post-lab Questions

1. To get an accurate measure of the energy content of the biodiesel fuel, all the heat released when it burns must be transferred to the water in the can. We never get it all. Where else can it go? As a result of these losses, do you expect your calculated heat content to be a little too low or a little too high when compared to the actual heat content of the biodiesel? Explain.

2. If your class prepared biodiesel from several different oils, prepare a small table listing the drip time and the heat content for each type of biodiesel. Now rank the fuels in order from lowest to highest energy content. Compare your ordered list to *Figure 11.9* in the *Chemistry in Context* textbook. What do you notice?

3. If you did Experiment 10, compare the heat content of your biodiesel to that of the hydrocarbons and alcohols of that experiment. Based strictly upon heat content per gram, which fuel is best? What factors besides high heat content must be taken into account when choosing a fuel?

4. The U.S. Department of Energy (DOE) has a website (http://www.afdc.energy.gov/afdc/) describing a number of alternative fuels, including biodiesel and ethanol. Go to the website and answer the following questions.

   a. Describe some of the benefits of biodiesel as listed on the DOE website (look under "Basics"). What further benefits do you think come from the production of used oil instead of new oil?

   b. What is ethanol? Give its chemical formula and describe how it is made.

   c. Describe some of the benefits of using ethanol as a fuel as listed on the DOE website.

   d. Much of the fuel ethanol produced in the United States is made from corn. Using the resources of the internet, describe the advantages and drawbacks to using corn for fuel. Cite your sources. What other feedstocks could be used to make ethanol that may offer advantages over corn?

Name _____   Date _____

Lab Partner _____   Lab Section _____

Data Sheet—Experiment 11

Part I: Preparation of Biodiesel

Type of oil used: _____

Record your observations during the transesterification reaction and isolation of biodiesel.

Part II: Viscosity

When a liquid flows, its molecules get tangled up. More tangling makes the liquid flow more slowly. Predict whether the viscosity of the oil will increase or decrease upon reaction to form biodiesel. Explain the basis for your prediction.

Perform the viscosity experiment and record the drip times in seconds for oil and biodiesel.

Liquid	Oil	Biodiesel
Trial 1		
Trial 2		
Average		

Which of these substances would flow more easily in an engine? _____

Comment on the validity of your hypothesis.

Part III: Temperature Effects

Observations for Oil:

Observations for Biodiesel:

Explain what might happen if you used biodiesel in your car during the winter.

Part IV: Heat Content of Biodiesel

Trial number	1	2
Mass of can + water (grams)		
Mass of empty can (grams)		
Mass of water (grams)		
Final temp. of water (°C)		
Initial temp. of water (°C)		
Temperature change (°C)		
Initial mass of burner (grams)		
Final mass of burner (grams)		
Mass of fuel burned (grams)		
Total heat absorbed by water (calories)		
Heat per gram of fuel (calories/gram)		
Average heat per gram of fuel (calories/gram)		

A Conductivity Detector for Ions
Can We Build our own Instruments?

INTRODUCTION

The modern chemical laboratory has an array of devices to assist chemists in gathering information. While modern labs tend to use commercial instrumentation, the equipment used by early scientists was hand-built by the individual scientists. In this exercise, you will build a simple electrical conductivity detector and then use it to test for the presence of ions in various liquids.[1] You can see a larger version of a conductivity detector in use in *Figure 5.17* of *Chemistry in Context*.

Background Information

Many chemical substances contain separate electrically charged particles called **ions.** (See Sections 5.6 and 5.8 in *Chemistry in Context* for more detail.) If ions are present, there must be equal amounts of both positively charged and negatively charged ions. As one example, when solid sodium chloride, NaCl, is dissolved in water, it breaks apart into Na^+ and Cl^- ions. Because these ions are free to move independently and because they carry electric charges, they can "conduct" electricity through the solution. This property provides a very simple and useful way to test for the presence of ions: If ions are present, the solution will conduct electricity. Furthermore, the magnitude of the conductance is directly proportional to the concentration of ionic substances dissolved in the liquid.

The device you will construct in this exercise is designed to indicate whether a solution can conduct electricity. When the probes are immersed in a conducting liquid, the light-emitting diode (LED) will light up. (Light-emitting diodes are solid state devices that emit light when electric current flows through them.) The detector is designed so that a conducting material completes the circuit and current flow causes the LED to emit visible light. The tester also works well for measuring the conductivity of solids: If the probes are touched to a solid and the LED lights up, then the solid conducts electricity.

Overview of the Experiment

1. Gather together all the parts for the detector.
1. Practice soldering if necessary.
1. Assemble and solder together the detector.
1. Test the detector to be sure it works.
1. Use the detector to test various liquids, solutions, and materials in the laboratory.
1. Take the detector with you to test materials outside of the laboratory.

---

[1] This design for a conductivity detector was originated by F. J. Gadeck, *J. Chem. Educ.*, **64**, 628 (1987).

Pre-lab Question
Do you think pure water will conduct electricity. Why or why not?

EXPERIMENTAL PROCEDURE

The electrical circuit diagram for the conductivity detector is shown in *Figure 12.1*.

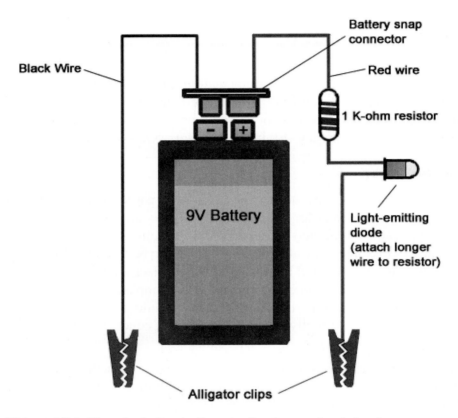

Figure 12.1 Electrical circuit diagram for the conductivity detector

I. Procedure for Soldering Wires Together

Solder is a low-melting mixture of several metals that is melted onto wires, which are then joined together as the solder solidifies, thus making an electrical connection. There are two options for soldering the wires in this exercise. Your instructor will indicate which one to use: (a) an electrically heated soldering iron or "gun" or (b) small strips of tape solder that are wrapped around the wires to be joined and then heated with a match or candle flame. The tape solder melts at a low temperature, thus making a soldering iron unnecessary.

Before assembling the detector, you should first practice soldering together some scrap pieces of wire as follows:

1. Using a wire stripper, strip off about 1 cm of the plastic coating from the ends of two wires. (There are several types of wire strippers. Your instructor will demonstrate the type that is available in your laboratory.)

2. Twist the bare ends of the wires together to make a mechanical connection.

3. Do ONE of the following

    a. If you are using a soldering iron or gun, obtain a strip of solder wire. Switch on the soldering iron or gun and hold it against the end of the solder wire until the solder begins to melt (to confirm that the soldering iron is hot). Then touch the hot tip of the soldering iron to the wires to be joined and simultaneously hold the end of the solder wire to the heated spot. Melted solder should flow smoothly over the hot wire junction. Remove the soldering iron and wait for the joint to cool. Check with the instructor to be sure you have a good joint.

    a. If you are using tape solder, cut a piece that is 0.5–1.0 cm long. Wrap the tape solder around the connected wires and then heat the tape solder gently with a match or candle. The tape solder should melt and flow over the wires, joining them together.

4. Repeat steps 1–3 until you can make a simple solder joint that seems solid.

II. Assembling the Detector

1. Gather the parts for the conductivity detector and be sure you can correctly identify each part. The parts are listed in the caption to *Figure 12.2*; you will also need two 12-inch lengths of thin plastic tubing and 12 inches of black electrical tape.

resistor   LED

2. Prepare the film canister by punching four holes in the cap, arranged as shown below. Punch the holes from the inside of the cap to avoid damaging the lip of the cap. If a suitable hole puncher is available, use it. Otherwise, use pliers to hold a nail in a flame until it is hot (not red hot). Then use the nail to melt suitable holes in the top of the plastic film container.

3. Look at *Figure 12.2* as you assemble your detector. Note the letters (A through G) that refer to each component in the figure. Push the two wires from the LED (part F) through the two closely spaced holes in the cap of the film canister (D).

4. Locate the longer wire of the LED. If thin plastic tubing is available, cut a length of it so that it covers all of the wire except for about 1 cm. This tubing, sometimes called "spaghetti", insulates the bare wires.

5. Twist the **long** wire of the LED and a wire from the resistor (E) together. (**Note:** If the wrong wire is attached to the resistor, the LED will be permanently damaged when current flows through the circuit.)

6. Solder the resistor and the LED together at this joint.

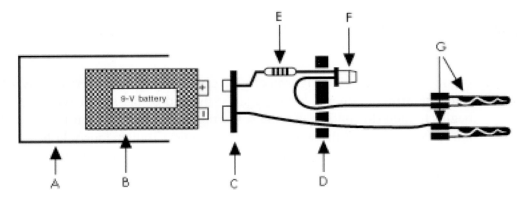

Figure 12.2 Diagram of the finished conductivity detector: (A) film canister, (B) 9-V battery, (C) battery clip, (D) canister cap, (E) resistor, (F) LED, and (G) alligator clips

7. If it is available, cut a length of thin tubing so that it covers all of the other resistor wire except for about 1 cm.

8. Solder the remaining resistor wire to the red wire from the battery clip (C).

9. Thread a wire through one of the remaining holes in the cap. Locate the other LED wire. If thin tubing is available, cut a length of it so that it covers all of the LED wire except for about 1 cm.

10. Solder the shorter wire of the LED to this wire.

11. Thread a second wire through the cap. Locate the black wire from the battery clip and solder it to the wire from this wire.

12. *Optional (recommended):* Attach alligator clips (G) to the ends of the two wires extending through the cap. First make a mechanical connection for each; then solder the connections.

13. Test your detector by connecting the battery clip to a 9-volt battery and touching the wires or the alligator clips together. The LED should light up.

14. If the detector does not work, check the entire system against the diagram in *Figure 12.1* to see if you have made a soldering error. Also check for loose connections. If you cannot find an error, check with your instructor.

15. If the detector works, package it in a black plastic film canister (*Figure 12.2*). Be careful not to pull apart any connections or have any bare wire connection touching any other bare connection. (Optional: After assembly, secure the top with black electrical tape.)

III. Using the Conductivity Detector to Test for Ions

If alligator clips are used, do not insert the metallic clips directly into the solutions. Instead, clip a short piece of wire (e.g., a partially unfolded paper clip) into each alligator clip. These wires can then be inserted into the solutions to be tested. Using wires in this fashion keeps the alligator clips out of the solutions and prevents the clips from becoming corroded.

To test a liquid sample, dip the wires attached to the alligator clips (or the wires themselves if no clips are used) into the liquid to be tested. The wires should dip in about a half an inch and should not touch each other. If the light goes on, it means that an electric current is flowing through the solution and that the solution contains ions. *It is important to rinse off the wires with pure water each time they are to be inserted into a new solution so as not to carry over contamination from one solution to the next.*

The tests that follow can be performed in small beakers, small test tubes, or in a plastic wellplate. Be sure to record your observations on the data sheet as you proceed. For each test, record whether you observe a bright light, a dim light, or no light.

1. Test the conductivity of (a) pure water (distilled or deionized) and (b) tap water. What can you conclude about the presence of ions in each?

2. Next, test a sample of salt water (1% solution of sodium chloride). Is the result what you expected?

3. Test a 1% solution of sodium hydroxide (formula: NaOH). Does it contain ions?

4. Hydrochloric acid is a water solution of HCl (which, by itself, consists of HCl molecules). Test a 1% solution of hydrochloric acid. What do you conclude? How would a chemist describe what is present in hydrochloric acid solutions?

5. Test a 1% solution of sugar in water. Does it contain ions?

6. Test a 10% solution of ethanol (ethyl alcohol) in water. Does it contain ions?

7. Using salt water, find out whether your conductivity tester is sensitive to the length of the wire that is immersed in the liquid being tested. Also find out whether the sensitivity of your detector is affected by the distance between the wires that are immersed in a liquid.

IV. Optional activities

1. Your instructor may provide additional solutions to test in the laboratory. If so, be sure to write down what the solutions are and what you observe.

2. Devise and carry out an investigation to determine how sensitive your detector is to the presence of ions in solution. Record your method and the results.

3. Now that you know how to use the conductivity tester, you may take it with you when you leave the lab and test at least five liquids that you encounter during the course of a day. Enter the data in the data sheet. Be prepared to report your observations at the next meeting of your class or lab section. Some suggestions for liquids to try include (a) tap water, (b) a stream or lake, (c) rainwater, (d) sweat, (e) urine, (f) foods such as coffee, tea, fruit juice, milk, soft drinks, a potato, a piece of apple or other fruit.

 CAUTION! When testing beverages and foods, always put a small amount of the substance in a separate clean container. *Never dip the tester into any food or beverage that you will later ingest.*

Post-lab Questions

1. Which of the tested liquids contained ions? Which ones did not contain ions?

2. Did the brightness of the light vary from solution to solution? Cite specific examples. What does a dim light tell us about the number of ions in a solution?

3. If you wanted to make a solution of an unknown solid substance to test it for the presence of ions, would it matter whether you used pure water or tap water to make the solution? Explain briefly.

4. If you were able to test some commercial products, read the labels on the containers of these liquids and attempt to determine what substances were responsible for the liquid's conductivity, if any.

5. The purity of ultrapure water is often measured with conductivity detectors. Based on what you have observed, why is this a good test for water purity? In what way might it be incomplete?

6. Describe the results you got when you varied the length of wire immersed and the distance between wires. Offer explanations for the behavior you observed.

7. What characteristic of ions allows us to test for them in this experiment?

8. What are some advantages and disadvantages of building your own lab equipment?

Name _____   Date _____

Lab Partner _____   Lab Section _____

Data Sheet—Experiment 12

Hypothesis

Before beginning the experiment, predict whether the samples you test will conduct electricity or not. State your reasons.

Tests of Liquids in the Lab

Liquid	Conductivity
Pure water (distilled or deionized)	
Tap water	
Salt water	
Sodium hydroxide solution	
Hydrochloric acid	
Sugar solution	
Ethanol solution	

Comment on your hypothesis.

Did you correctly predict the conductivity of the solutions? In cases where you did not, how have your experimental results caused you to revise your ideas about the structures of the compounds in these solutions.

What was the effect of varying the length of wire immersed in the solution?

What was the effect of varying the distance between the wires?

Optional activities:

Liquids Tested Outside the Lab

Liquid	Conductivity

Sensitivity of the detector
Write out the procedure you used to test the sensitivity of your detector. Include your results and conclusions.

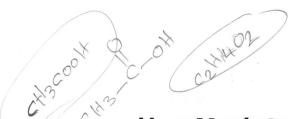

How Much Acid is in Food?
Analysis of Vinegar

INTRODUCTION

Whether you prefer to sip a soda or a refreshing glass of lemonade, acids are an important component of our daily food. As Chapter 6 in *Chemistry in Context* points out, acids in our air are also a serious environmental problem. In this laboratory assignment you will analyze vinegar, a common ingredient in pickles and salad dressing for its acid content. You will use a common laboratory technique called titration. Since later laboratory assignments will also use the titration method, the background information and the discussion of the procedure given here are quite detailed.

Background Information

As described in Sections 6.1 and 6.2 of *Chemistry in Context*, solutions of **acids** in water contain excess H^+ ions, while solutions of **bases** contain excess OH^- ions. Vinegar is a dilute solution of acetic acid, $HC_2H_3O_2$, in water.

A characteristic property of acids and bases is that they react with each other. Thus, for example, acetic acid in vinegar will react with sodium hydroxide (a base) to produce water and sodium acetate:

$$\text{acetic acid} \quad + \quad \text{sodium hydroxide} \quad \rightarrow \quad \text{water} \; + \; \text{sodium acetate}$$
$$HC_2H_3O_2 \quad + \quad NaOH \quad \rightarrow \quad H_2O \; + \; NaC_2H_3O_2$$
$$\text{(an acid)} \qquad\qquad \text{(a base)}$$

This type of reaction is the basis for the **titration method of analysis** for acids. In a titration, a known quantity of acid solution is measured out; then a solution of base (usually sodium hydroxide, NaOH) is added slowly until just enough has been added to react with all of the acid. If one knows the volume of NaOH added and its concentration, it is possible to calculate how much acid must have been present to react with the NaOH. Finally, if the volume of the acid solution is known, it is possible to calculate the concentration of the acid. Concentrations are expressed as **molarity** (M), which is defined (Section 5.5 in the text) as the number of moles of a substance in exactly 1 liter of solution.

In a titration, an **indicator** is added. This is a chemical substance that has one color in the presence of excess H^+ ions and a different color when there are excess OH^- ions. The change of color signals the end of the titration, i.e., the point where just enough NaOH has been added to completely react with all of the acid. The indicator used in this experiment is *phenolphthalein*, which is colorless in acid and red in base. To do a titration, a drop of the indicator is added to a measured amount of acid. Then NaOH solution is added, one drop at a time, until one drop of NaOH changes the solution from colorless to pink. This is the "end point" of the titration. Indicators are intensely colored so only one drop is needed for a titration.

For accurate calculations, the exact volumes of the acid and base must be known. If we assume that the plastic pipets used in the experiment deliver drops of a uniform size, then instead of measuring volumes directly, you can count drops of each solution to obtain the volumes of acid and of base used. From the number of drops for each reactant and the concentration of sodium hydroxide, it is possible to calculate the concentration of acid in the vinegar solution.

Overview of the Experiment

1. Obtain a 24-well wellplate, plastic transfer pipets, and the necessary solutions.
2. Practice dispensing drops evenly.
3. Practice titrating vinegar with sodium hydroxide solution.
4. Titrate samples of vinegar to determine their acetic acid content.
5. Calculate the molarity and percent acidity of the acetic acid in the vinegar.

Pre-lab Question

Look up and draw the structure of acetic acid, showing how the atoms are bonded together. What functional group (see Section 10.3 in *Chemistry in Context*) does it contain?

EXPERIMENTAL PROCEDURE

I. Preliminary work

1. Obtain a plastic wellplate with 24 wells and a small stirring rod or toothpick. Place the wellplate on a sheet of white paper so that color changes can be seen easily. Your wellplate probably has letters and numbers to identify the wells. If not, refer to the diagram in *Figure 1.2* in Experiment 1. These labels are used in the following instructions.

2. Label three plastic pipets as follows: vinegar, NaOH, indicator.

3. Fill the "vinegar" pipet from a bottle of vinegar and record the brand of vinegar.

4. Fill the "NaOH" pipet from the supply bottle. Record on the data sheet the exact concentration of the NaOH solution (look on the label of the supply bottle).

 CAUTION! Sodium hydroxide solution is corrosive and can cause serious skin and eye damage. Therefore, you MUST wear eye protection at all times and avoid getting any of the solution on your skin. In case of skin contact, rinse immediately with plenty of water and notify your instructor.

5. Partially fill the "indicator" pipet with phenolphthalein solution.

II. Pipet Practice

Fill the unlabeled fourth plastic pipet with water and try dispensing drops into a well of the wellplate. Your goal is to be able to confidently add a known number of drops to the bottom of the well, one drop at a time. Squeeze the bulb gently, while the pipet is held vertically and directly over the center of the well. You may find it helpful to use two hands to steady the pipet. Plastic transfer pipets easily acquire air bubbles in their stems that lead to frustrating "half-drops" and "quarter-drops" that introduce errors. It is best to to reserve one well (e.g., D6) for solution waste. During a titration, "half-drops" can be added to the waste well rather than the titration well. When you can confidently add a known number of drops to a well, proceed to the practice titration.

III. Practice Titration

1. Add 10 drops of vinegar to well A1 in the wellplate.

2. Then add 1 drop of indicator. The mixture should be colorless.

3. Start adding NaOH, <u>one drop at a time</u>, (counting the drops), with gentle stirring. Carefully observe what happens. As each drop is added, a slight pink color may be visible. This color should disappear upon stirring. After enough drops have been added, the solution in the well will turn pink and stay pink. Make a note of how many drops were used. Then add a few more drops of NaOH, one drop at a time, with stirring. The color will likely become darker red as the solution becomes more and more basic.

IV. Measurement of the Acetic Acid Content of Vinegar

In doing titrations, your goal is to catch the point where <u>one drop</u> of NaOH produces the first permanent color change.

1. Carefully add <u>20 drops</u> of vinegar (count them) to each of six wells (B2, B3, etc.). Make sure that the drops fall directly to the bottom of the well and are not "trapped" along the side. If you make a mistake, simply start again in another well. Record the position numbers for the wells you are using (B2, B3, etc.) on the data sheet.

2. Add <u>1 drop</u> of indicator to each well.

3. Start adding NaOH, <u>one drop at a time</u>, to the first well as you stir the solution. As you get close to the stopping point, you will find that the momentary pink color persists longer after each drop is added. Stop when one drop of NaOH produces a permanent pink or red color. Record the number of drops you used on the data sheet. If you make a mistake, skip this trial and go on to the next one.

4. Repeat this titration for each of the other five wells. With a little practice, the titrations go very fast. Be sure to record the number of drops used for each well on the data sheet.

   Hint: Once you know approximately how many drops of NaOH are needed, you can quickly add about 2/3 of that amount. For example, if the first titration required 15 drops of NaOH, then for the next trial, you could quickly add about 10 drops of NaOH and then add the remaining drops slowly.

   Note: If you miss the end point, or lose count of drops or are uncertain of the result for any other reason, it is good scientific practice **not** to erase the data. Simply make a note in the table as to what you think went wrong, draw a line through that row, and do another titration. All scientists make mistakes when doing lab work! Results crossed out will not affect your grade.

5. You should now have six results. They ought to agree with each other within a few drops. If they do not, consult your instructor.

6. **Optional extension:** Analyze another kind of vinegar. First, rinse the vinegar pipet with the new sample by twice filling the pipet and emptying it into the appropriate waste container. Use a new set of wells in the wellplate for the new sample.

Clean-up

Your instructor will specify where to dispose of the contents of the wellplates and whether to wash or dispose of the plastic pipets. Wash the wellplate thoroughly and allow it to dry.

V. Calculations

NOTE: A calculator will be quite helpful for these calculations.

The logic behind the calculations for a titration analysis is a little tricky. Even if you do not understand the explanation completely, the actual calculations are quite easy.

The balanced chemical equation for the reaction of acetic acid with sodium hydroxide is

$$HC_2H_3O_2 + NaOH \rightarrow NaC_2H_3O_2 + H_2O.$$

This equation shows that *1 mole of acetic acid reacts with 1 mole of sodium hydroxide*. Thus, for the titration of vinegar, the number of moles of NaOH added from the pipet is exactly equal to the number of moles of acetic acid (from vinegar) in the well. Expressed mathematically, we can say that

$$\text{moles of NaOH added from the pipet} = \text{moles of } HC_2\ HC_2H_3O_2\ H_3O_2 \text{ in the well} \quad (1)$$

The number of moles of NaOH in any sample of NaOH solution can be calculated from the molarity of the NaOH (expressed as moles/liter) and the volume (expressed as liters), using the following equation:

$$\text{moles of NaOH} = \frac{\text{moles of NaOH}}{1 \text{ liter of NaOH solution}} \times \text{liters of NaOH solution} \quad (2)$$

A similar equation can be written for the moles of acetic acid in the vinegar sample.

$$\text{moles of } HC_2H_3O_2 = \frac{\text{moles of } HC_2H_3O_2}{1 \text{ liter of vinegar}} \times \text{liters of vinegar} \quad (3)$$

By substituting into the equation 1, we can write the following equation:

$$\frac{\text{moles of NaOH}}{1 \text{ liter of NaOH solution}} \times \text{liters of NaOH} = \frac{\text{moles of } HC_2H_3O_2}{1 \text{ liter of vinegar}} \times \text{liters of vinegar} \quad (4)$$

Or, writing this more succinctly,

$$\text{molarity of NaOH} \times \text{liters of NaOH} = \text{molarity of } HC_2H_3O_2 \times \text{liters of vinegar} \quad (5)$$

This equation can be rearranged to put it into a more useful form:

$$\text{molarity of } HC_2H_3O_2 = \text{molarity of NaOH} \times \frac{\text{liters of NaOH}}{\text{liters of vinegar}} \quad (6)$$

There is one small problem: You don't know the actual volumes in liters or milliliters; you know only the number of drops of each solution that you used. If we assume that all of the drops from both pipets have the same volume, then the ratio of volumes expressed as drops should be the same as the ratio of volumes expressed as liters.

$$\frac{\text{liters of NaOH}}{\text{liters of vinegar}} = \frac{\text{drops of NaOH}}{\text{drops of vinegar}} \quad (7)$$

This finally gives us equation (8) for calculating the molarity of acetic acid:

$$\text{molarity of } HC_2H_3O_2 = \text{molarity of NaOH} \times \frac{\text{drops of NaOH}}{\text{drops of vinegar}} \quad (8)$$

1. To do the calculations, first look carefully at the six or more titration results in the data table. Does any one result seem way out of line from the others? If so, you probably did something wrong, and it is legitimate to omit that result. Simply write "omit" in the comments box. Using all of the results that you think are valid, calculate the average number of drops of NaOH and record this on the data sheet.

2. Next, calculate the molarity of acetic acid in the vinegar, using equation (8). Multiply the molarity of the NaOH by the average number of drops of NaOH and then divide by 20, the number of drops of vinegar used.

3. Labels on vinegar bottles usually specify the *percent acidity*, which is defined as the number of grams of acetic acid per 100 mL of vinegar. One mole of acetic acid weighs 60 grams; therefore, multiplying the molarity (moles/liter) by 60 will give the number of grams of acetic acid per liter and then dividing by 10 will give the number of grams per 100 mL (since 100 mL = 1/10 of a liter). Using this information, calculate the percent acidity for your vinegar sample and record it on the data sheet.

An example will make this clearer. Suppose you found that the acetic acid concentration was 0.55 mole per liter of vinegar. Then the calculations would be as follows:

4.2%

$$\frac{0.55 \text{ mol } HC_2H_3O_2}{1 \text{ liter of vinegar}} \times \frac{60 \text{ g } HC_2H_3O_2}{1 \text{ mol } HC_2H_3O_2} = \frac{33 \text{ g } HC_2H_3O_2}{1 \text{ liter of vinegar}}$$

$$\frac{33 \text{ g } HC_2H_3O_2}{1 \text{ liter of vinegar}} = \frac{33 \text{ g } HC_2H_3O_2}{1000 \text{ mL of vinegar}} = \frac{3.3 \text{ g } HC_2H_3O_2}{100 \text{ mL of vinegar}} = 3.3\% \text{ acetic acid}$$

Report your results in whatever manner is specified by your instructor.

Post-lab Questions

1. During the experiment, you were cautioned about making sure that all drops of solutions were at the bottom of the well, not along the sides. Why?

2. Suppose that during the titration, you added sufficient NaOH to change the color of the indicator to a darker pink. How would this change the results? Would your calculated molarity be higher or lower than the actual molarity? Explain.

3. What is the advantage of doing several trials, especially since you are doing exactly the same thing each time?

4. Could you have done the titration "backwards," i.e., start with 20 drops of NaOH, add phenolphthalein indicator, and then add the vinegar one drop at a time? Explain. What would you observe?

5. Can you think of ways to improve this method of analysis? What changes might be possible to give you more confidence in the accuracy of the results?

6. Why do you suppose that the acetic acid content in vinegar is reported as percent acidity rather than as molarity?

7. Suppose you were asked to determine the exact acetic acid concentration in an acetic acid solution that was about 50% acid. How would you change the procedure you used to determine the acetic acid content of vinegar so it could be used to solve this problem?

8. Titration can be used to determine the acid content of any food or drink. List three that you believe contain acid. (Hint: Acidic foods often taste tart.) Titrations of highly colored substances require some modification of our procedure. Why would a highly colored food present problems for this procedure?

What Makes Water Hard?
Measurement of Water Hardness

INTRODUCTION

As you know from Section 5.5 of *Chemistry in Context*, and may have also discovered in Experiment 12, drinking water contains many dissolved ions. Water with high levels of calcium ions (Ca^{2+}) and magnesium ions (Mg^{2+}) is known as **hard water**. These ions are responsible for an assortment of problems ranging from aesthetic ones, such as bathtub rings and soap scum, to more critical ones such as plugged steam lines and damaged hot water heaters. Calcium ions get into water if the water comes in contact with limestone (calcium carbonate); magnesium ions likewise enter water when ground water passes through minerals that contain magnesium ions.

In this experiment, you will analyze water samples for the combined concentrations of calcium and magnesium ions known as "total hardness." You will use a titration method of analysis, which is described in the text. The titration method is fast, easy, and one of the most widely used analytical techniques in many areas of chemistry and biology.

Background Information

The analysis for water hardness is based on the chemical reaction of Ca^{2+} (and Mg^{2+}) with an ion called dihydrogenethylenediaminetetraacetate, $C_{0}H_{12}N_2O_8^{2-}$. Since the name and formula are complicated, it is almost always called simply "EDTA" (and often written as " H_2EDTA^{2-} "). The reaction involves one calcium (or magnesium) ion reacting with one H_2EDTA^{2-} ion. (Note that in this reaction, one Ca^{2+} ion replaces two H^+ ions.)

$$Ca^{2+} + H_2EDTA^{2-} \rightarrow Ca(EDTA)^{2-} + 2\,H^+$$

The analytical procedure is a **titration.** A known amount of water is measured out, and EDTA solution is added slowly until just enough has been added to completely react with all of the calcium and magnesium ions in the sample. In order to know when enough EDTA solution has been added, a small amount of an **indicator** called calmagite is used. At pH 10 (an alkaline solution), the indicator by itself has a deep blue color, but in the presence of metal ions, it changes to red. Thus, if a drop of indicator is added to a solution containing calcium and magnesium ions, the solution will have a reddish color. When enough EDTA has been added to react with all of the calcium and magnesium, the solution will turn blue.

The strategy of the experiment will be to first use an EDTA solution to titrate a reference calcium solution of known concentration and then use the same EDTA solution to titrate an unknown water sample. By comparison of the two sets of results, it is easy to calculate the hardness of the unknown sample.

For simplicity, the amounts of Ca^{2+} and Mg^{2+} that are present in the water sample will be lumped together. Furthermore, the common practice in water hardness is to report the results as *milligrams of CaCO3 per liter of water*. This does not mean that solid calcium carbonate is actually present, but rather that this amount of solid calcium carbonate could be formed from the amount of Ca^{2+} (and Mg^{2+}) present in the water.

Overview of the Experiment

1. Test the indicator color change.
2. Do 3 or 4 titrations of water containing a <u>known</u> concentration of calcium ions.
3. Calculate the calibration factor for the EDTA solution.
4. Do 3 titrations for each of several water samples containing an <u>unknown</u> concentration of calcium and magnesium ions.
5. Calculate the water hardness for the unknown samples.

Pre-lab Question

Look at a periodic table and identify the locations of magnesium and calcium. From the discussion of atomic structure in Chapter 1 of *Chemistry in Context*, explain why these two elements may react in a similar manner to each other in this experiment.

EXPERIMENTAL PROCEDURE

I. Preliminary Work

1. Obtain 2 small test tubes that are <u>clean</u> and <u>dry</u> and a rack to hold them. Fill them about 1/3 full with the following solutions. Be sure to label them so that you know which one is which.
   a. EDTA solution
   b. Reference solution of known hardness   (equivalent to 0.500 mg of CaCO3 per mL of water)

2. Obtain 4 plastic graduated-stem pipets that are clean and dry.
   a. Label one as "buffer" and fill it 1/3 full with "buffer solution, pH 10."
   b. Label another as "indicator" and fill it 1/3 full with "calmagite indicator."
   c. Label a third pipet as "EDTA" and place it in the EDTA test tube.
   d. Label a fourth pipet as "Ca reference" and place it in the reference solution.
   Be sure that all pipets are labeled so that you will not get them mixed up.

3. Obtain a 24-well plastic wellplate and a short stirring rod made of glass or plastic. Place the wellplate on a piece of white paper (to aid in seeing the color changes). Note the letters and numbers on the wellplate (or refer to the diagram in *Figure 1.2* in Experiment 1). Have the data page of this experiment ready to record numbers as you obtain them.

4. <u>Pipet practice</u>. (If you have done a previous titration experiment with this kind of equipment, such as Experiment 13, you can skip this step.)

   Take an empty graduated-stem plastic pipet, fill it with water, and try dispensing drops into a well of the wellplate. Your goal is to be able to confidently add a known number of drops to the bottom of the well, one drop at a time. Squeeze the bulb gently while the pipet is held vertically and directly over the center of a well. You may find it helpful to use two hands to steady the pipet. These pipets easily acquire air bubbles in their stems, which lead to frustrating "half-drops" and "quarter-drops" that introduce errors. The best solution is to reserve one well for solution waste. During a titration, "half-drops" are not added to the titration well or counted. They are simply added to the waste well. When you can confidently add a known number of drops to a well, proceed to the practice titration.

II. Titrations of the Reference Solution

1. Test the color change of the indicator. To well A1, add 8–10 drops of pure water. Then add 2 drops of pH 10 buffer and 1 drop of calmagite indicator. Stir with a clean stirring rod. The solution should be a bright, clear blue. If not, add 1 drop of the EDTA solution to produce a blue color. This is the color you will be looking for at the end of each titration. It should be *pure blue* with no trace of red. (If not, check with your instructor before proceeding further.)

2. Rinse and fill a 1-mL graduated-stem plastic pipet with the calcium reference solution. Practice until you can fill it *exactly* to the 1-mL line. (The best method for doing this is to gradually squeeze air out of the bulb, release the pressure, and observe how high the solution rises. Repeat until the solution just rises to the 1-mL mark.) Dispense exactly 1 mL of this solution into each of four wells in the wellplate (e.g., wells A2–A5).

3. To the first well, add 2 drops of buffer plus 1 drop of indicator (in that order). Then start adding EDTA solution slowly, one drop at a time, counting drops and stirring after each addition. Continue until the color starts to change from red to purple. Wait a few moments to see if the color continues to change (the reaction is slow); then *slowly* add additional drops, stirring after each one, until the color becomes pure blue. Keep track of the total number of drops added. When you reach the end point, enter in the table the number of drops used.

4. Repeat the same procedure with the samples in the other three wells, titrating one sample at a time. Remember to add 2 drops of buffer and 1 drop of indicator (always in that order) before each titration.

5. Do additional titrations if you are uncertain of any of the four titrations (for example, if you lost any drops of solution or think that you went past the end point or lost count). With a little practice, the titrations can be done quite rapidly. (The slowest part of this particular titration may be waiting for the chemicals to react completely.)

III. Titrations of Unknown Water Samples

Obtain one or more water samples to be analyzed, as directed by your instructor. These might be the tap water in your laboratory or in your dormitory or home. They might be samples from a nearby lake or river. They might be from a well or a spring.

Note: Some water samples – especially tap water in some buildings – may not give a good color change to pure blue. This is caused by traces of iron or copper in the water. If you find you have such a sample, check with the instructor before proceeding further. It may be necessary to titrate to a purple color rather than a pure blue.

1. Obtain a clean 1-mL graduated-stem pipet. Label it for water samples.

2. Carefully dispense 1 mL of a water sample into a clean well in the wellplate—e.g., B1.

3. Add 2 drops of buffer followed by 1 drop of calmagite indicator. Proceed to titrate with the EDTA solution in exactly the same way you did in Part I. Add the EDTA one drop at a time, keeping track of the number of drops added. Stir after each addition, watch for the start of the color change and then continue slowly until one drop turns the solution pure blue. Because this is an unknown solution, you have no idea the first time whether it will take more or fewer drops of EDTA than the reference solution did; thus, you will need to do the first titration slowly. Record the total numbers of drops of each solution used in the data table.

4. Do two or three additional titrations for this sample. Depending on how much EDTA was used in the first trial, you may want to use either a larger or smaller volume of the unknown sample for the remaining titrations. (If you do use a different volume of water, be sure to record this in the data table.) Remember to add the buffer and indicator each time. Record the numbers in the data table.

IV. Optional titrations

1. You may be curious to analyze additional water samples or your instructor may assign another sample to be analyzed. Be sure to record any pertinent sample information on your data sheet prior to beginning your analysis. Remember to rinse the sample pipet thoroughly with each new sample before measuring out 1-mL portions of that sample, or, if possible, use a new, clean pipet.

2. Take about 10–20 mL of water that is relatively "hard" and allow it to run slowly through a small tube containing beads of an ion exchange resin. Collect some of the softened water in a clean test tube and titrate 1-mL portions of the water using the same procedure as above. Is the hardness of the water different? Give a simple chemical explanation of what happened in the tube.

3. A simple distillation apparatus may be set up in the laboratory. See Figure 5.30 in *Chemistry in Context,* for a similar apparatus. Tap water or other hard water will be placed in the heated flask. Obtain a few milliliters of the distilled water in a clean test tube and titrate 1-mL portions of this water using the same procedure as above. How does the hardness of distilled water compare with the starting material?

4. Your instructor may ask you to analyze some water samples for calcium and magnesium using modern analytical instrumentation. The instrument is first calibrated using solutions of known concentration and then unknown solutions are analyzed. You will be given detailed instructions in how to use the instrument and do the accompanying calculations. You can then compare the instrumental results with those obtained by the titration procedure described previously.

Clean-up

Dispose of solutions and pipets in appropriate containers as instructed. Wash your wellplate and allow it to drain.

V. Calculations

1. Look carefully at the data from each of the sets of titrations to see whether the results show consistency or whether any one result in a given set should be eliminated because it appears to be an "outlier." If so, make a note beside it in your notebook.

2. Calculate the average number of drops of EDTA used for each set of titrations, omitting any results that you think are not valid.

3. In order to do the calculations, you need to know the "hardness" of the reference solution. This should be written on the bottle or posted in the lab. Ideally it will be close to 500 mg/liter, which is the same as 0.500 mg/mL. (Remember that hardness is normally expressed as mg of $CaCO_3$ per liter.)

4. For the first set of titrations (with the reference solution), divide the milligrams of hardness per milliliter by the average number of drops of EDTA used.

$$\frac{\text{mg hardness in ref. solution}}{1 \text{ mL ref. solution}} \times \frac{1 \text{ mL of ref. solution}}{\text{avg. drops of EDTA}} = \frac{\text{mg of hardness}}{1 \text{ drop of EDTA}} \qquad (1)$$

This gives the mg of hardness corresponding to 1 drop of EDTA solution. This is the <u>calibration factor</u>. It is unique for the particular EDTA solution and your equipment.

5. For the analysis of an unknown water sample, multiply the average number of drops of EDTA used by the calibration factor to obtain the mg of hardness per mL of water.

$$\frac{\text{drops of EDTA}}{1 \text{ mL sample}} \times \frac{\text{mg of hardness}}{1 \text{ drop of EDTA}} = \frac{\text{mg of hardness}}{1 \text{ mL sample}} \qquad (2)$$

6. Finally, convert this into milligrams of hardness *per liter* of water. This is the normal way of reporting water hardness.

$$\frac{\text{mg of hardness}}{1 \text{ L sample}} = \frac{\text{mg of hardness}}{1 \text{ mL sample}} \times \frac{1000 \text{ mL}}{1 \text{ L}} \qquad (3)$$

7. Rate the hardness of each water sample analyzed, using the following classification.

Description	Concentration
Very hard	Over 300 mg/L of calcium carbonate
Hard	150 to 300 mg/L of calcium carbonate
Moderately hard	50 to 150 mg/L of calcium carbonate
Soft	0 to 50 mg/L of calcium carbonate

VI. Sharing Your Results with Other Students

Your instructor will provide details on how to share your results with other students. If everyone has analyzed the same sample, results can be assembled and compared. If the class was divided up to analyze various samples, you should record what other students found for the other samples.

Post-lab Questions

1. Based on your own data and/or class data, how would you classify the hardness of your local tap water?

2. If your class analyzed other water samples (lake, river, well water, etc.), did you find significant differences in hardness? If so, suggest possible explanations.

3. If you investigated the use of an ion exchange resin, how effectively did it soften the water? What were the levels of hardness before and after treatment?

4. If you analyzed a sample of water from a distillation apparatus in the lab, what did you conclude about its effectiveness in removing calcium and magnesium ions?

5. In these titrations, there may be an uncertainty of at least one drop in identifying the exact end point. If you use 20 drops of EDTA solution in a titration, what percent uncertainty in the hardness is contributed by adding one extra drop?

$$\% \text{ error} = \frac{\text{no. of drops you are in error}}{\text{no. of drops used}} \times 100\%$$

6. Is very soft water the same as very pure water? Explain.

7. Write a summary paragraph describing the causes of hard water, the symptoms you might notice if your household has hard water, and ways to soften water.

Name _____   Date _____

Lab Partner _____   Lab Section _____

Data Sheet—Experiment 14

Part II: Titrations of the Reference Solution

Hardness of the reference solution: _____ mg $CaCO_3$ per 1 mL

Trial	#1	#2	#3	#4	#5
mL of reference sol'n	1.00	1.00	1.00	1.00	
Drops of EDTA					

Average drops of EDTA sol'n = _____ per 1 mL of reference solution

Calibration factor = _____ mg of hardness per 1 drop of EDTA   **(Eq. 1)**

Part III: Titration of Water Samples

Description of Sample #1: _____

Titration	#1	#2	#3	#4	#5
mL of water sample	1.00				
Drops of EDTA					

(Extra columns are provided in case they are needed)

Average drops of EDTA sol'n: _____ per 1 mL of water

mg of hardness per 1 mL of water: _____   **(Eq. 2)**

mg of hardness per liter of water: _____   **(Eq. 3)**

Rating of water hardness: _____

Space is provided on the reverse side of this sheet for up to 3 more water samples or studies of water softening. For any additional analyses, use this same format and record the information on an additional sheet of paper, which should be attached to this data sheet when submitted to the instructor.

Description of Sample #2: _____

Titration	#1	#2	#3	#4	#5
mL of water sample	1.00				
Drops of EDTA					

(Extra columns are provided in case they are needed)

Average drops of EDTA sol'n: _____ per 1 mL of water

mg of hardness per 1 mL of water:_____   **(Eq. 2)**

mg of hardness per liter of water: _____   **(Eq. 3)**

Rating of water hardness: _____

Description of Sample #3: _____

Titration	#1	#2	#3	#4	#5
mL of water sample	1.00				
Drops of EDTA					

Average drops of EDTA sol'n: _____ per 1 mL of water

mg of hardness per 1 mL of water:_____   **(Eq. 2)**

mg of hardness per liter of water: _____   **(Eq. 3)**

Rating of water hardness: _____

Description of Sample #4: _____

Titration	#1	#2	#3	#4	#5
mL of water sample	1.00				
Drops of EDTA					

Average drops of EDTA sol'n: _____ per 1 mL of water

mg of hardness per 1 mL of water:_____   **(Eq. 2)**

mg of hardness per liter of water: _____   **(Eq. 3)**

Rating of water hardness: _____

How Does Human Activity Affect Water Purity? Measurement of Chloride in River Water

INTRODUCTION

Water contains many dissolved ions, as you know from Sections 5.6 and 5.8 of *Chemistry in Context*. The chloride ion is a negatively charged ion found in water and sewage. The chloride content of natural (unpolluted) surface waters depends on the geology of the area. In areas where surface water normally has very little chloride, a higher chloride concentration implies a source from human activity. In this experiment, you will measure the chloride concentration in water samples taken from streams or rivers in your local area and also, if possible, samples from your local wastewater treatment plant. By dividing up the class to analyze different samples, it will be possible to collect a large quantity of data in a short time. You will then use the results to draw conclusions about the human impact on local surface waters.

Note that chlor<u>ide</u>, Cl⁻, (an ion) is quite different from chlor<u>ine</u>, Cl_2, (a gaseous molecule). Chlorine is added as a disinfectant to drinking water in small amounts (less than 1 ppm) and to swimming pools in somewhat larger amounts (about 5 ppm). Chlorine will produce some chloride as it reacts, but in most cases, it is not a major source of chloride.

Background Information

Solutions containing chloride ions (Cl⁻) will react with silver nitrate ($AgNO_3$) to form an insoluble white compound, silver chloride (AgCl). (Other ions that are present in the water do not participate in this reaction.)

$$AgNO_3(aq) + Cl^- \text{ (in water)} \rightarrow AgCl(s) + NO_3^-(aq) \qquad \textbf{(1)}$$
$$\text{(white)}$$

This reaction is the basis for the titration method of analysis for chloride ions. In a **titration**, a known volume of a water sample containing chloride ions is measured out, and then a solution of $AgNO_3$ is added slowly until just enough has been added to react with all of the chloride in the sample. If the volume of $AgNO_3$ added and its concentration are known, it is possible to calculate how much chloride must have been present to react with all the $AgNO_3$. Finally, if the volume of the water sample is known, the concentration of chloride ion in the sample can be calculated. Concentrations are expressed as **molarity** (M), defined as *moles of substance per liter of solution*, but in this analysis (as in many other analyses), the final results will be expressed in different units.

In order to know when enough silver nitrate has been added, a small amount of an **indicator** is added. A solution of sodium chromate (Na_2CrO_4) is used, in which chromate ions (CrO_4^{2-}) serve as the indicator. Chromate ions are yellow, but they react with silver ions to form a red precipitate of silver chromate (*Equation 2*, next page). As silver nitrate is added, the chloride is precipitated as white silver chloride (*Equation 1*). After all of the chloride has all been removed, silver ions (Ag⁺)

will react with CrO_4^{2-} to form a red insoluble precipitate of silver chromate, Ag_2CrO_4 (*Equation 2*). The appearance of this red precipitate signals the end of the titration.

$$2 \, Ag^+(aq) \; + \; CrO_4^{2-}(aq) \rightarrow Ag_2CrO_4(s) \;\; \text{(red precipitate)} \qquad \textbf{(2)}$$
$$\text{(red)}$$

To do a titration, *1 drop* of indicator is added to a measured volume of water, then $AgNO_3$ is added, one drop at a time. You will notice your yellow solution turn orange as the red precipitate forms.

For greatest accuracy, the exact volume of each solution used must be known. However, if we assume that the plastic transfer pipets deliver drops of a uniform size, then instead of measuring exact volumes, you can count drops of each solution to obtain the volume of water sample and the volume of silver nitrate solution. If the number of drops of each reactant is known and the molarity of $AgNO_3$ is known, it is possible to calculate the concentration of chloride ions in the water sample.

Overview of the Experiment

1. Collect water samples at various locations.
2. Practice doing chloride titrations with tap water and silver nitrate.
3. Calculate the concentration of chloride in the tap water sample.
4. Titrate the collected water samples with silver nitrate.
5. Calculate the concentration of chloride in the collected samples.
6. Compare class data for different water samples.
7. Draw conclusions about human impact on the local aquatic environment.

Pre-lab Question
Define the term *carcinogen*.

EXPERIMENTAL PROCEDURE

I. Collect Water Samples

An important aspect of the experiment is sample collection and choice of sampling locations. Before the lab class, your instructor will assign a sample location for each pair of students (or will arrange to have samples collected and made available in the laboratory). If a major river or stream flows through your town or city, then water samples should be collected from the river upstream of the local city, from the river in the middle of the city, and downstream of the city far enough away that any sewage treatment plant effluent has freely mixed into the river. Samples should also be collected from the sewage treatment plant itself: both the effluent and the influent (which should be aerated because of its high suspended solid content). Your class also needs to have samples from the municipal drinking water sources, such as reservoirs or lakes. Finally, it is helpful, for comparison purposes, to collect samples from sources that are "clean," such as ponds or small streams, that do not have any likely human impact. If you collected your own sample(s), be sure to record accurately the location from which you sampled the water and label the sample(s).

II. Preliminary Work

1. Obtain a plastic wellplate with 24 wells and a small stirring rod. Place the wellplate on a sheet of white paper so that color changes can be seen easily.

2. Obtain two graduated-stem plastic pipets (or other matched-style pipets). Label one "water sample" and the other "silver nitrate" (or "AgNO3"). *Labeling is crucial because it is easy to get them mixed up!*

3. Using a clean *dry* beaker or other container, obtain about 10 mL of silver nitrate solution. Label the container and record the exact concentration (given on the label of the supply bottle) on the data sheet.

     CAUTION! Silver nitrate will stain your skin black, so be careful not to get it on your hands. The stain is harmless and will wear off in a few days, but in the meantime, its appearance is unattractive.

4. Obtain one or more water samples for analysis. These may be samples you have collected yourself (as described above) or samples provided by the instructor. You will start by analyzing tap water.

III. Practice Titration

1. Obtain some tap water in another beaker. Using the "water sample" pipet, add 10 drops of this tap water to well A1 in the wellplate.

2. Add 1 drop of sodium chromate indicator. The mixture should be pale yellow-green.

     CAUTION! Sodium chromate is considered to be a carcinogen. Avoid contact with your skin.

3. Start adding AgNO$_3$ solution, <u>one drop at a time</u> (counting the drops), with gentle stirring. Carefully observe what happens. As each drop is added, you will observe a cloudy appearance as solid silver chloride is formed; then a reddish color may appear that will disappear on stirring. After a number of drops have been added, the mixture in the well will have an orange tint. This first appearance of a uniform tint signals the end point. Make a note of how many drops were used. Then add 1 or 2 more drops of silver nitrate with stirring. The color will become darker orange or red due to formation of more Ag$_2$CrO$_4$.

4. Finally, add more tap water, with stirring, until the color reverts back to a milky yellow-green. Save this mixture for color comparison.

IV. Analyze Tap Water

In doing titrations, your goal is to catch the point where one drop of AgNO$_3$ produces the first permanent color change. It is not a dramatic color change but rather a slight orange tint. This is the "end point" where all of the chloride has been used up by reaction with the AgNO$_3$.

1. Carefully add 20 drops of tap water (count them) to each of four wells (B1, B2, etc.). Make sure that the drops fall directly to the bottom of the well and are not "trapped" along the side. If you make a mistake, simply start again in another well. Record the position numbers (B1, B2, etc.) for the wells you are using on your data sheet.

Hints for good pipet technique: (1) Best results are obtained when the pipet is held vertically while dispensing drops. If you hold it at an angle, try to always use approximately the same angle. (2) It is useful to set aside one well in the wellplate (e.g. D6) as a "waste container." If there is an air bubble or a partial drop forming, it can be discarded in the waste container.

2. Add 1 drop of indicator to each well.

3. Start adding $AgNO_3$, <u>one drop at a time</u>, to the first well as you stir the solution. Stop when one drop of $AgNO_3$ produces a permanent orange color. Record the number of drops you used on the data sheet. If you make a mistake, skip this trial and go on to the next one.

Hint: The gradual color change from yellow to pale orange may be hard to recognize. If so, it is helpful to use the mixture in well A1 for comparison. You are looking for the first change away from that color.

4. Repeat this titration for each of the other 3 wells. With a little practice, the titrations go very fast. Record the number of drops you used for each well on the data sheet.

Hint: Once you know approximately how many drops of $AgNO_3$ are needed, you can quickly add about 2/3 of that amount. For example, if the first titration required 15 drops of $AgNO_3$, then for the next trial, you could quickly add about 10 drops of $AgNO_3$ and then add the remaining drops slowly.

Note: If you missed the end point or lost count of drops or are uncertain of the result for any other reason, it is good scientific practice **not** to erase the data. Simply make a note in the table as to what you think went wrong, draw a diagonal line through that column, and do another titration. Results crossed out will not affect your grade for the experiment.

5. If desired, titrate an additional one or two samples of tap water.

V. Analyze Collected Water Samples

1. Rinse the pipet and beaker used for tap water, first with distilled or deionized water and then with a new sample to be analyzed.

2. Do a trial titration with 20 drops of the new sample in one well (e.g., C1). Add 1 drop of sodium chromate indicator, then titrate with silver nitrate as before. Depending on the nature of the sample, it may have very little chloride, requiring only a few drops of silver nitrate, or the chloride concentration may be very high, requiring a large quantity of silver nitrate. Record the number of drops used.

3. Depending on the results from this first titration, you may decide to use more than 20 drops or less than 20 drops of water. Dispense samples of the desired size into three more wells.

4. **Optional extension**: If time permits, you may be asked to analyze one or more additional water samples. If additional samples are done, record your results on a separate sheet of paper and attach it to the data sheet.

VI. Clean Up

Dump the contents of the wellplate into a designated waste container. Rinse the wellplate with tap water, then with distilled or deionized water if available. Leave it upside down on a paper towel to drain. Your instructor will indicate whether the pipets are to be discarded or saved.

VII. Calculations

You should now have at least four results for tap water and four results for at least one other water sample. The same calculation procedure is used for both sets of data. First, some background for the calculations is presented.

The balanced chemical equation for the reaction of chloride ion with $AgNO_3$,

$$Cl^- \text{ (in water)} + AgNO_3(aq) \rightarrow AgCl(s) + NO_3^-(aq) \tag{1}$$

shows that 1 mole of chloride reacts with 1 mole of $AgNO_3$. Thus, for the titration of a water sample, the number of moles of silver nitrate added equals the number of moles of chloride ions (in the water sample) that was in the well.

$$\text{moles of chloride in the well} = \text{moles of } AgNO_3 \text{ added} \tag{3}$$

For the left side of equation (3), the moles of chloride = (molarity of chloride) $\times$ (vol. of water), and for the right side, the moles of $AgNO_3$ = (molarity of $AgNO_3$) $\times$ (vol. of $AgNO_3$).

Substituting into equation (3) gives

$$\text{(molarity of chloride)} \times \text{(vol. of water)} = \text{(molarity of } AgNO_3\text{)} \times \text{(vol. of } AgNO_3\text{)} \tag{4}$$

If we assume that the drops are all the same volume, then we can say that

$$\text{(molarity of chloride)} \times \text{(drops used)} = \text{(molarity of } AgNO_3\text{)} \times \text{(drops used)} \tag{5}$$

Rearranging equation (5) gives the final working equation:

$$\text{molarity of chloride} = \text{molarity of } AgNO_3 \times \frac{\text{drops of } AgNO_3}{\text{drops of water sample}} \tag{6}$$

1. To do the calculations, first look carefully at the four titration volumes for tap water. Does any one result seem way out of line from the others? If so, you probably did something wrong, and it is legitimate to omit that result. Simply write "omit" underneath it. Using all of the results that you think are valid, calculate the average number of drops of $AgNO_3$ used to titrate the tap water and record this on the data sheet.

2. Use equation (6) above to calculate the molarity of chloride in the tap water.

3. Finally, convert the answer to milligrams of chloride per liter of water (mg/L), which is the same as parts per million. To do this, multiply the molarity of chloride by 35,500 mg of Cl per 1 mole of Cl. (One mole of chloride has a mass of 35.5 g, which is the same as 35,500 mg.)

Example: Suppose that you found the chloride concentration was 0.0012 mole/liter. Then the calculations would be as follows:

$$\frac{0.0012 \text{ mol chloride}}{1 \text{ L water}} \times \frac{35,500 \text{ mg chloride}}{1 \text{ mol chloride}} = \frac{43 \text{ mg chloride}}{1 \text{ L water}} = 43 \text{ ppm} \tag{7}$$

4. Repeat the calculations in steps 1–3 for your collected water sample(s).

VIII. Interpretation of the Class Data

When class data are assembled, the class will examine the results to see what patterns emerge. In particular, you will try to assess how much effect the human use of water has on water quality in the local environment. In many places, the water is used by a city and passed on to other communities downstream. It is useful to ask what has been done to it in the process. Finally, your class will try to answer the question, "Where does the extra chloride come from?"

Report your results in whatever manner is specified by your instructor.

Post-lab Questions

1. During the experiment, you were cautioned about making sure that all drops of solutions were at the bottom of the well, not along the sides. Why?

2. Suppose that during the titrations you consistently added sufficient $AgNO_3$ to change the color of the indicator to a darker orange-red. Would this change the results? Would your measured chloride concentrations be higher or lower than the actual concentrations? Explain.

3. What is the advantage of doing three or more trials, especially since you are doing exactly the same thing each time?

4. After examining the assembled class data, what can you conclude about the impact of your city on the local river or stream?

5. List some of the major nonindustrial sources of chloride in natural waters. Which ones do you think are important in your area?

6. Based on what you have learned in this experiment, do you consider chloride a serious water pollutant? Why or why not?

Name _____   Date _____

Lab Partner _____   Lab Section _____

Data Sheet—Experiment 15

Molarity of the $AgNO_3$ solution: _____ mole/liter

Part IV: Tap Water

Trial number	Well number	Drops of water	Drops of $AgNO_3$	Comments
1				
2				
3				
4				
5				
6				

Note that extra rows are provided in case they are needed.

Average number of drops of $AgNO_3$: _____ (per 20 drops of tap water)

Concentration of chloride: _____ mole/liter   **(eq. 6)**

Concentration of chloride: _____ mg/liter (or ppm)   **(eq. 7)**

Part V: Collected Water Sample

Description of the water sample: _____

Hypothesis

Based on the type of sample and the location collected, do you think this sample will have a higher or lower chloride concentration than tap water? Why?

Trial number	Well number	Drops of water	Drops of $AgNO_3$	Comments
1				
2				
3				
4				
5				
6				

Note that extra rows are provided in case they are needed.

Average number of drops of $AgNO_3$: _____

Concentration of chloride: _____ mole/liter   **(eq. 6)**

Concentration of chloride: _____ mg/liter (or ppm)   **(eq. 7)**

Conclusion

Was your hypothesis correct? Explain.

What's in My Bottled Water?

INTRODUCTION

Chapter 5 in *Chemistry in Context* discusses bottled water. In this experiment, you will analyze a sample of bottled water to measure the concentrations of several of the dissolved substances likely to be present in the water. You will use techniques that you may have used in previous experiments, notably titrations with colored indicators and use of a pH meter.

This study will be a class project, involving several different analyses and a variety of bottled water samples. Teams of four or more students will take one bottled-water sample and divide up the work for the analyses. At the end, class results will be assembled for comparison and discussion.

Background Information

The sources and composition of bottled water are discussed in *Chemistry in Context*, Chapter 5. Bottled water is rarely, if ever, pure water. Because it often comes from wells or springs, bottled-water frequently has a relatively high mineral content. In particular, it is likely to have calcium and magnesium ions, the primary sources of **water hardness**. If calcium and magnesium are high, there is usually also a high concentration of bicarbonate ion, HCO_3^-, which accounts for the **alkalinity** or **acid neutralizing capacity** of the water. In addition, there will be varying amounts of other soluble ionic substances.

The pH of water must be close to neutral (pH 7) in order for the water to be safe for human consumption, but it will be slightly on the alkaline side (i.e., above pH 7) if there is substantial bicarbonate ion present.

Another way to characterize bottled-water samples (or any other water samples) is the total amount of dissolved material, often referred to as **total dissolved solids (TDS)**. This consists mostly of ionic substances, since water is an excellent solvent for ions. Some bottled-water labels indicate the TDS concentration.

For easy comparisons, all experimental results (except pH) will be converted to parts per million (milligrams of a substance per liter of water).

Overview of the Experiment (Not all steps may be done)

1. Measure the pH of the water.
2. Measure the concentration of calcium and magnesium by titration.
3. Measure the bicarbonate ion concentration by titration.
4. Measure the chloride ion concentration by titration.
5. Measure the total dissolved solids in the water.
6. Assemble the analytical information and compare results for different samples.

Pre-lab Question

Identify two reasons why people drink bottled water, and two problems with the consumption of bottled water.

EXPERIMENTAL PROCEDURE

This experiment includes several different analyses. Your instructor will specify which part(s) you should do and how the class will be divided.

I. Measurement of the pH

This should be done following the procedure described in Experiment 18. If you are already familiar with pH measurements and if a pH meter has already been calibrated, the actual measurement will take only a few minutes. If you are doing pH measurements for the first time, you need to read the background information and follow the step-by-step instructions carefully.

An alternate, fast, but less precise, way to measure pH is to use pH test strips.

II. Concentration of Calcium and Magnesium

This can be done by a titration procedure described in Experiment 14. Water samples are titrated with a solution of EDTA. The color change of an indicator (calmagite) shows when enough EDTA has been added to exactly react with all of the calcium and magnesium.

> **Note:** If an EDTA solution of known concentration is available, there is no need to do the initial titrations of a reference solution as specified in Experiment 14. In that case, a specific number of drops of water should be used rather than 1 mL. The calculation of calcium + magnesium molarity then becomes
>
> $$\text{molarity of Ca} + \text{Mg} = \text{molarity of EDTA} \times \frac{\text{drops of EDTA}}{\text{drops of water}} \quad \textbf{(1)}$$
>
> For an explanation of the logic behind this calculation, refer to Experiment 13 or Experiment 15.

Calculations: The analysis measured the *combined* concentration of Ca and Mg. Since calcium is almost always present in much larger concentration than magnesium, we will simplify the calculation by assuming it is all calcium. Calculate the concentration as parts per million, ppm, by one of the following methods: (a) If you calculated the molarity of Ca + Mg by equation 1 above, multiply this molarity by 40,000 mg/mole, the molar mass of Ca expressed as milligrams. (b) If you used the calculation procedure in Experiment 14, multiply the "hardness" by 0.40 to convert from mg $CaCO_3$ per L to mg Ca per L. (Remember that this is actually Ca + Mg.)

III. Concentration of Chloride

This is a titration analysis to determine the concentration of chloride ion (Cl^-), described in detail in Experiment 15. Samples of water are titrated with a silver nitrate solution of known concentration. An indicator (sodium chromate) is added to show when enough silver nitrate has been added to exactly react with all of the chloride.

Calculations: Do these exactly as specified in Experiment 15, reporting the final result as ppm of chloride in water.

IV. Concentration of Bicarbonate

Analysis for bicarbonate is not described elsewhere in this laboratory manual, so more detailed instructions are provided here. The procedure is straightforward, using the same basic titration technique employed in Experiments 13, 14, and 15. In this case, bicarbonate ion (HCO_3^-) in the water is titrated with hydrochloric acid solution (HCl). The equation for the chemical reaction is

$$HCO_3^- \ (aq) \ + \ HCl \ (aq) \ \rightarrow \ H_2O \ + \ CO_2 \ (g) \ + \ Cl^- \ (aq)$$

A colored indicator, methyl orange, will change color from yellow to orange when an excess of HCl is present so that the solution becomes acidic. (More precisely, the indicator changes color when the pH drops below 4.)

A. Get Organized for Titrations

1. Obtain a plastic wellplate with 24 wells and a small stirring rod. Place the wellplate on a sheet of white paper so that color changes can be seen easily.

2. Using a clean, *dry* beaker or other container, obtain about 10 mL of hydrochloric acid solution (HCl). Label the container and record on the data sheet the exact concentration given on the label of the supply bottle.

3. Pour some of the bottled water into another clean, *dry* beaker or other container and label it.

4. Obtain two graduated-stem plastic pipets (or other matched-style pipets). Label one "water sample" and the other "HCl" and place them in the corresponding solution. *Labeling is crucial because it is easy to get them mixed up!*

5. If you have *not* previously done titrations with plastic disposable pipets, you should practice dispensing drops into the wellpate (e.g., somewhere in row D) until you can reliably add one drop at a time. Some useful advice on pipet technique is given in Experiment 13 under "Pipet Practice."

B. Titration of Bottled Water

1. Using the "water sample" pipet, add 10 drops of this water to well A1 in the wellplate.

2. Add 1 drop of methyl orange indicator. The mixture should be yellow.

3. Start adding HCl solution, <u>one drop at a time</u> (counting the drops), with gentle stirring after each addition. Carefully observe what happens. Watch for the *first color change* from yellow to orange that persists after the mixture is stirred. Record the number of drops used.

4. To aid in recognizing the color change for subsequent titrations, add a few more drops of bottled water to well A1. The color should go back to yellow. Save this for color comparison. *You will be looking for the first color change away from this yellow.*

5. From the result of the first titration, decide how much water to use for subsequent titrations so as to require a convenient amount of HCl. If possible, you should aim for at least 10 drops and no more than 30 drops of HCl.

6. Carefully add the desired number of drops of water to wells A2 through A4.

7. Add 1 or 2 drops of methyl orange indicator to each.

8. Titrate the sample in well A2, adding HCl, one drop at a time, to the first color change that persists after stirring. (Use well A1 as a reference color.) Record the number of drops in the data table.

9. Titrate the next two samples in the same manner. If the 3 or 4 results do not show adequate consistency, do one or two additional titrations.

Calculations: The logic of the calculations is the same as that shown in Experiment 15 for chloride ion. First, calculate the molarity of bicarbonate ion by the following equation:

$$\text{molarity of } HCO_3^- = \text{molarity of } HCl \times \frac{\text{no. of drops of HCl}}{\text{no. of drops of water sample}} \quad \textbf{(2)}$$

To convert to ppm, multiply the molarity of HCO_3^- by 61,000 mg HCO_3^- per mole of HCO_3^-.

V. Concentration of Total Dissolved Solids

1. Check to be sure that a large hot plate has been turned on and is ready to use.

2. Obtain two clean, dry 50-mL beakers. Label them with numbers and your initials. Weigh each and record the mass to the nearest milligram (0.001 gram). *First check to be sure the balance reads 0.000 g with nothing on the pan.*

3. Using a graduated cylinder, carefully measure out 20 mL of bottled water and add it to the weighed beakers.

4. Place the beakers on a hot plate, adjusted so that water will boil gently. (Be sure you can identify your sample since other samples will probably be on the same hot plate.) Continue heating until the water has *completely* evaporated from the beakers.

5. Remove the beakers (*caution: use tongs*) and allow them to cool. Inspect carefully to be sure they are totally dry, with no hint of moisture remaining in the bottom or on the sides of the beakers.

6. Reweigh the beakers and record the masses. *First check to be sure the balance reads 0.000 g with nothing on the pan.*

Calculations: Subtract to find the mass of solid in a beaker, then calculate the ppm of TDS as follows. First convert the volume of water used to liters (e.g., 20 mL is 0.020 L).

$$\text{ppm of TDS} = \frac{\text{g of solid x } 1,000 \text{ mg / g}}{\text{vol of water (in L)}} \quad \textbf{(3)}$$

VI. (Optional) Use an Instrumental Procedure to Measure Calcium and Magnesium.

The titration procedure in Part II is a fast and relatively simple way to measure the *combined* concentrations of calcium and magnesium. A more elegant method can be used to measure each ion separately. This requires an expensive, partially automated instrument that is typical of equipment used in modern chemical analysis laboratories. If this step is assigned, your instructor will provide detailed instructions for the particular equipment being used. The general strategy is to measure absorption or emission of radiation for a series of standard reference solutions of each metal, then measure the water sample and compare it with the standards. A calibration graph is needed, which is either prepared by hand or done electronically by a computer within the instrument. One advantage of this kind of analysis is that, once the calibration has been done, many samples can be analyzed very rapidly.

VII. Clean-up

When finished with all analyses, follow all instructions regarding waste disposal. Put solutions into appropriate waste containers. Your instructor will tell you what to do with the plastic pipets. The remaining glass and plastic items should be rinsed thoroughly with deionized or distilled water and left upside down to drain.

VIII. Comparison and Reporting of Results

Your team should assemble the analyses for your particular water sample, then post them or report them in some fashion to the rest of the class. Your instructor will specify how to report your results and compare them to results from other groups. You may be able to compare brands of bottled water, and in some cases compare your results to what is reported on the label for the water.

Post-lab Questions

1. Two negatively charged ions were measured, chloride and bicarbonate. Which is present in higher concentration in the bottled-water sample? To answer this question from a chemical viewpoint, you will need to compare *molarities*.

2. Using your results, what is the ratio of the bicarbonate *molarity* to the combined *molarity* of calcium + magnesium? If the calcium/magnesium and bicarbonate all came from limestone dissolved in the water, then the bicarbonate molarity should be exactly twice that of the calcium + magnesium molarity. How closely do your results agree with this prediction?

3. The pH of water for human consumption should be close to neutral. How closely does your bottled-water sample compare with this goal?

4. Check the labels on the bottled waters that you studied. Were they merely treated, or were they purified in some way? If so, list the method(s) used. What was the specific goal of the treatment? See Sections 5.11 and 5.12 of your textbook for help.

5. **Optional** (This question is more challenging.) The text shows a label from one brand of bottled water. For the positive ions that are listed, calculate the total *molarity* of all positive ion charges. (Keep in mind that an ion such as calcium, Ca^{+2}, has two positive charges.) Do the same thing for the negative ions. (If sulfate, SO_4^{2-}, is present, it has two negative charges.) How well does the *sum* of the positive charge molarity equal the *sum* of the negative charge molarity?

Name _____   Date _____

Partner(s)/team members _____   Lab Section _____

Data Sheet—Experiment 16

Note: For the parts done by other members of your team, note this and simply enter the final results.

Description of bottle-water sample _____

Part I: pH _____

Part II: Calcium and Magnesium

(Use the data table in Experiment 14 <u>or</u> the following table)

Trial #	1	2	3	4	5
Well number					
Drops of water					
Drops of EDTA					
Molarity of Ca + Mg (equation 1)					

(extra columns are provided if needed)

Average molarity: _____ mole/liter

Calculated ppm of Ca + Mg _____ mg/L

Part III: Chloride

Molarity of the $AgNO_3$ solution: _____ mole/liter

Trial #	1	2	3	4	5
Well number					
Drops of water					
Drops of $AgNO_3$					
Molarity of Cl (eq 5, Experiment 15)					

(extra columns are provided if needed)

Average molarity: _____ mole/liter

Calculated ppm of chloride _____ mg/L

Part IV: Bicarbonate

Molarity of the HCl solution: _____ mole/liter

Trial #	1	2	3	4	5
Well number					
Drops of water					
Drops of HCl					
Molarity of HCO_3^- (equation 2)					

(extra columns are provided if needed)

Average molarity: _____ mole/liter

Calculated ppm of HCO_3^- _____ mg/L

Part V: Total Dissolved Solids

	Trial 1	Trial 2
Volume of water used, mL		
Mass of beaker, g		
Mass of beaker + solid, g		
Mass of solid, g		
TDS, mg/L (equation 3)		

Does Acid Reign?
Reactions of Acids with Common Substances

INTRODUCTION

Acids are an important category of chemical substances. They are found in many places, including certain foods, our bodies, and frequently in rain or snow. (See Chapter 6 in *Chemistry in Context* for more detail.) Acids undergo chemical reactions with a great variety of substances. Some of these reactions are desirable, while others can be quite damaging. For instance, the acids found in rain or snow can have destructive effects on various materials.

In this experiment, you will investigate the reactions of three common acids (hydrochloric acid, sulfuric acid, and nitric acid). You will study their reactions with some familiar substances: marble (a building material), four common metals (zinc, copper, nickel, and aluminum), and a sample of protein (egg white) that represents living matters. In order to make observations in a very limited time, you will use acid solutions that are far more concentrated than those found in rain or snow. Still, the same reactions do occur with acid rain, but on a vastly slower time scale.

Background Information

Acids are substances that contain hydrogen ions (H^+) or that react with water to form hydrogen ions. Chapter 6 in *Chemistry in Context* provides more detail on acids and discusses the environmental consequences of the acids found in rain.

Most metals react with acids to produce hydrogen (H_2), a colorless gas, and metal ions. However, as you will see, not all acids and metals react in exactly the same way. Careful observation will also show that the rate of reaction of acids with materials varies with concentration. Building materials vary widely in their reactivity with acid. Some, like marble (a form of calcium carbonate), react readily with acids. Others, such as granite, do not visibly react at all. The reaction of proteins with acids is unlike that of any of the other materials. In aqueous solutions, protein molecules have a preferred three-dimensional shape that is pH dependent. Addition of an acid changes the pH and the protein shape so that the protein becomes less soluble. In addition to causing solubility changes, some acids also chemically react with the protein molecule itself.

In this experiment, the concentrations of acids are expressed as molarity, which is discussed in *Chemistry in Context* Section 5.5. The **molarity** of a substance in solution (often designated by M) is defined as the number of moles of that substance in 1 liter of solution.

Overview of the Experiment
1. Test the reactions of three acids (HCl, H_2SO_4, HNO_3) with marble chips (calcium carbonate).
2. Test the reactions of these acids with several metals (zinc, copper, nickel, aluminum).
3. Test the effect of these acids on egg white (a sample of protein).
4. Find out whether a change in acid concentration affects the speed of the reactions.

Pre-lab Question

Locate zinc, copper, nickel and aluminum on a periodic table. Do you expect their reactions with acid to be similar or different? Why?

EXPERIMENTAL PROCEDURE

 STOP! You should always wear eye protection in the laboratory, but this is especially important when working with acids. Safety glasses are absolutely essential to protect your eyes from any splashes.

I. Preparation of Metal Samples

The tests with metals will work best when the metal surface has been freshly cleaned to remove any corrosion or coating. Therefore, you should obtain the following sets of metal pieces and clean at least one surface of each piece with either sandpaper, emery paper, or steel wool.

1. Zinc: These can be either small, thin strips of zinc or galvanized nails (which are made of iron coated with zinc). When clean, the surface should be bright and shiny.

2. Copper: These may be very short pieces of heavy-gauge copper wire or small pellets of copper shot. Be sure the surface is bright and shiny.

3. Nickel: Fresh, shiny paper clips are usually made of iron or steel with a coating of nickel. They should not need any cleaning

4. Aluminum. These can be obtained most conveniently by cutting small strips out of an empty beverage can and then cleaning one surface to expose fresh metal.

II. Preparation of the Wellplate

1. Place a clean, dry 24-well wellplate on a sheet of white paper. Label the rows and columns on the paper according to the diagram below. If possible, also have available a dark surface on which to place the wellplate (e.g., a bench top or dark paper). Some tests show up better when viewed against a dark background rather than a white background.

2. Add one dropper-full (about 20–30 drops) of 6 M hydrochloric acid (HCl) to each of the wells in row A. The wells should be 1/4 to 1/2 filled.

 CAUTION! Be careful not to spill acids on your clothing, books, papers, or yourself. If you do spill any acid on your skin, wash it off promptly with large amounts of water. Continue to run water on the affected area for several minutes. Notify your instructor immediately in the event of a spill on any person or workspace.

3. Similarly, add one dropper-full of 6 M sulfuric acid (H_2SO_4) to each well in row B.

4. Row C can be filled by 10-fold dilution of the 6 M sulfuric acid: e.g., 18 drops of water + 2 drops of 6 M acid.

5. Fill all wells in row D with 6 M nitric acid (HNO_3).

6. Have available a paper towel on which to lay test objects after they have been used. Also have available a plastic wash bottle filled with pure water.

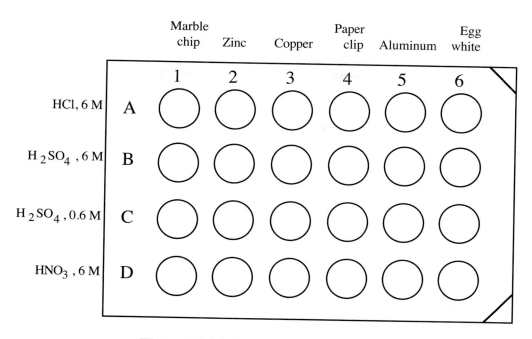

Figure 17.1 Diagram of the wellplate setup

III. Investigation of Reactions

Note: This experiment involves many tests and many observations. Careful observation is essential because the reactions of acids are so varied. *It is important to record your observations as you make them, rather than waiting until the end.*

CAUTION! Nitric acid reacts with most metals to produce poisonous gases. Therefore, all of the tests with metals and nitric acid (step 7 below) MUST be done in a fume hood. Alternatively, you can do the entire experiment in the fume hood if space is available.

1. Put a small marble chip in each well in column 1. Observe carefully what happens and record all of your observations. Do you see evidence of chemical reactions? Pay particular attention to differences between the acids and to differences in reaction rates.

    (**Optional extension:** Test the reaction of these acids with two other familiar forms of calcium carbonate: chalk and eggshell.)

2. Put zinc strips into the hydrochloric and sulfuric acid wells in column 2 (but *not* into the nitric acid well). Try to leave a portion of metal sticking out of the acid so that you have a "handle" with which to remove it when the test is completed. Observe the reaction of the zinc with each acid solution and record your observations. Pay particular attention to differences in the reactions and to differences in reaction rates.

3. Place clean pieces of copper in the hydrochloric and sulfuric acid wells in column 3 (but not into the nitric acid well). Observe the reaction of the copper with each acid solution and record your observations. Pay particular attention to differences in the reactions and to differences in reaction rates.

4. Put small, shiny paper clips into the hydrochloric and sulfuric acid wells in column 4 (but not into the nitric acid well). Observe the reaction of the paper clip with each acid solution and record your observations. Pay particular attention to differences in the reactions and to differences in reaction rates.

5. Put a small piece of aluminum in each well in column 5. Observe the reaction of the aluminum with each acid solution and record your observations. Pay particular attention to differences in the reactions and to differences in reaction rates.

6. Put several drops of egg white (protein) in the wells in column 6. Observe the reaction of the egg white with each acid solution and record your observations. Pay particular attention to differences in the reactions and to differences in reaction rates.

7. Finally, take your wellplate and fresh samples of the four metals to a fume hood. When the wellplate is in the fume hood, put small pieces of zinc, copper, aluminum, and nickel into the appropriate nitric acid wells in row D. Observe the reaction of the metals with the nitric acid solution and record your observations. Pay particular attention to differences in the reactions and to differences in reaction rates.

 CAUTION! Do not put your face close to the wellplate because a poisonous gas may be produced. When finished, either leave your wellplate in the fume hood until the evolution of gas has stopped or else lift out the pieces of metal before removing the wellplate from the hood.

IV. Clean-up

When you have completed the experiment, empty the contents of your wellplate into the waste container provided and follow any other instructions for disposal and cleaning. If there are any pieces of marble or metal remaining in your wellplate, put them in the appropriate containers. Finally, rinse the wellplate several times with tap water. Your instructor may give further instructions on what to do with the wellplate.

Post-lab Questions

1. What general conclusions can you draw from your observations? Do acids react with these materials? Did the reactions all appear to be the same? If there were differences, was there any pattern to the behavior?

2. Marble is calcium carbonate, $CaCO_3$. Its reaction with hydrochloric acid is

$$CaCO_3 + 2\,HCl \rightarrow CaCl_2 + CO_2 + H_2O$$

What <u>gas</u> was produced? What reactions do you think occurred when marble was exposed to the other acids? Do these reactions have anything in common? In light of your observations, speculate on the damage to marble statues and buildings caused by acid rain.

3. The reactivity of marble with sulfuric acid seems curiously out of line with all of the other observations. If you observed carefully, you probably saw a brief burst of gas bubbles, which quickly stopped. This is because one product of the reaction, calcium sulfate, $CaSO_4$, is very insoluble and provides a protective coating on the marble. This slows the reaction dramatically. Does this mean that acid rain containing sulfuric acid should have little or no effect on statues and buildings?

4. Do zinc, nickel, iron, and aluminum react with acids?   Based on your observations, how effective are zinc coatings and nickel coatings as a protection for iron against acid rain.

5. The gas produced in the reactions of metals with acids is H_2. Write a chemical equation for the reaction of zinc (Zn) with hydrochloric acid. One product is zinc chloride, $ZnCl_2$.

6. Copper has been important since ancient times as a building material used on roofs of buildings. On the basis of your observations, is copper a good material for this purpose? (Note: You may have observed a green color of roofs on public buildings or monuments such as the Statue of Liberty. This is a form of copper carbonate, formed by slow reaction of copper with nitric acid plus the carbon dioxide in the atmosphere.)

7. What generalization can you make about the effect of changing the concentration of an acid? Suppose the acid concentrations were reduced 100-fold, 1,000-fold, or even more?

8. Aluminum reacted differently from the other metals. What differences did you observe? Aluminum easily forms a very unreactive oxide coating. How can this account for the observed differences in aluminum's reactivity?

9. What reactions did you observe between the egg white and the acids?   Stomach acid is approximately 0.16 M HCl. Comment on what effects stomach acid might have on any protein you eat. Would you expect the reaction of stomach acid with protein in food to be faster or slower than the reaction you observed between egg white and 6 M HCl? Think about the effect of concentration on reactions.

10. Does acid reign? Explain the pun in the title of this experiment.

Notes

Name _____   Date _____

Lab Partner _____   Lab Section _____

Data Sheet—Experiment 17

Acid	Observations with marble
6 M HCl	
6 M H_2SO_4	
0.6 M H_2SO_4	
6 M HNO_3	

Acid	Observations with zinc
6 M HCl	
6 M H_2SO_4	
0.6 M H_2SO_4	
6 M HNO_3	

Acid	Observations with copper
6 M HCl	
6 M H_2SO_4	
0.6 M H_2SO_4	
6 M HNO_3	

Acid	Observations with nickel
6 M HCl	
6 M H$_2$SO$_4$	
0.6 M H$_2$SO$_4$	
6 M HNO$_3$	

Acid	Observations with aluminum
6 M HCl	
6 M H$_2$SO$_4$	
0.6 M H$_2$SO$_4$	
6 M HNO$_3$	

Acid	Observations with egg white (protein)
6 M HCl	
6 M H$_2$SO$_4$	
0.6 M H$_2$SO$_4$	
6 M HNO$_3$	

Which Common Materials are Acids or Bases?

INTRODUCTION

The acidity or basicity of substances can be determined by measuring their pH. In this experiment, you will learn how to calibrate and operate a modern pH meter. You will measure the pH values for a variety of substances in water, including some pure chemicals, some rain and water samples, and various foods and household products. You may also have the opportunity to try some simple "dip sticks," which are disposable pH test strips.

Because this experiment includes measuring the pH of common foods and household substances, you may wish to bring a few substances of your own choosing to lab so you can measure each one's pH.

Background Information

The pH scale is a convenient way of expressing hydrogen ion concentration in solutions. Section 6.6 in the text discusses pH and the pH scale. pH is defined as the negative of the logarithm of the hydrogen ion molarity (M_H^+). (See Appendix 3 of the text for more detail.) In equation form, this is

$$pH = -\log M_{H+} \tag{1}$$

Modern pH meters are wonderfully easy to use and generally reliable. But the science behind their operation is far more complicated. They rely on an *extremely* small electric current flowing through the solution and the meter, and thus they are highly susceptible to a variety of interferences. Probably the most important considerations in making accurate pH measurements are (a) *careful* adjustment and rechecking of the meter, (b) *thorough* rinsing of the electrode with pure water between each measurement, and (c) care in handling the electrode because it is very fragile (and also expensive).

Although the following directions describe the use of an electronic instrument for measurement of pH, there is a much simpler method for measuring approximate pH values using special pH test paper strips. These have been impregnated with colored dyes that change color depending on the pH. Most of this experiment can be done satisfactorily with pH test strips instead of a pH meter. One disadvantage of test strips is that most of them are reliable only to the nearest whole-number pH unit. For some brands of test strips, the pH can be estimated to within a few tenths of a pH unit.

Overview of the Experiment
    1. Learn the operation of a pH meter and calibrate the meter.
    2. Measure the pH of pure acid and base solutions.
    3. Measure the pH of various foods and household products.
    4. Measure the pH of tap water, rainwater, and surface water.
    5. Measure the pH of water containing two atmospheric gases: CO_2 and (optional) SO_2.

Pre-lab Question
Predict whether each household substance listed in Part III below is acidic or basic, and explain your reasoning.

EXPERIMENTAL PROCEDURE

Note: The following directions assume that you will be using a pH meter. If you are using pH test strips instead of a pH meter, you can skip Part I and go directly to the tests in Parts II to V.

I. Operation and Calibration of a pH Meter

1. Obtain a pH meter with an attached probe called a *pH electrode.* Your instructor will explain the type of pH meter to be used and arrangements for sharing meters between students. If the pH meter requires AC voltage, be sure it is plugged in and warmed up.

2. Obtain two small beakers containing reference solutions labeled "pH 7.00" and "pH 4.00."

3. Learn how to operate the pH meter, including the functions of whatever knobs or buttons it has. Most meters have three knobs: TEMPERATURE, SLOPE, and CALIBRATE (or STANDARDIZE).

4. Hold the electrode over a waste container and rinse the electrode thoroughly with pure water from the wash bottle. Blot the end of the electrode *gently* with a tissue to remove most of the water.

5. Insert the electrode into the pH 7.00 calibrating solution, stir gently for a few moments, and observe the reading on the meter. Use the CALIBRATE knob to adjust the meter until it displays 7.00. You may need to keep stirring. The meter may not settle down to display *exactly* 7.00, but it should be within the range 6.9 to 7.1.

6. Rinse the electrode thoroughly with pure water, blot the end of the electrode *gently* with a tissue, then insert the electrode into the pH 4.00 calibrating solution, stir and observe until the reading is steady. If necessary, adjust the SLOPE (or TEMP) control until it reads very close to 4.00.

7. Recommended: Repeat with both pH 7.00 and 4.00 (rinsing and blotting each time the electrode is moved) to be sure the readings are steady.

II. Measure the pH of Chemical Solutions of Known Concentration

1. Remove the electrode and rinse it <u>very thoroughly</u> with pure water, since it is now coated with a concentrated pH 4 or pH 7 solution, which could make subsequent readings erroneous. Blot the end of the electrode *gently* with a tissue to remove most of the water.

2. Immerse the electrode into a 0.0001 M solution of HCl (in a test tube or small beaker), stir the solution for a few moments, and record the pH. Calculate the predicted pH for this solution using equation 1 and record that also.

3. Remove the electrode, rinse it well, blot with a tissue, then immerse the electrode into 0.001 M HCl, stir, and record the pH along with the predicted pH.

4. Rinse the electrode and measure the pH of 0.01 M HCl.

5. Finally, switch to an alkaline (basic) solution, as follows. Rinse the electrode thoroughly and measure the pH of a 0.001 M solution of sodium hydroxide, NaOH. Compare your result with the calculated pH for this solution. (Review how to calculate pH for a base.)

III. pHs of Foods and Household Substances

Now that you know how to measure pH, you should measure the pH of some common substances, using samples provided by the instructor and samples brought by students. In each case, put a <u>small</u> amount in one of the wells of a plastic wellplate or in a small test tube (or in a very small beaker). Use only enough so that the bulb end of the electrode is immersed. If in doubt about this, check with the instructor.

Some possible samples to test include the following substances. You should test 6–8 substances or whatever number your instructor specifies.

    a. Vinegar   d. Soft drinks
    b. Lemon juice   e. Coffee or tea
    c. Fruit juices

<u>The following substances should be diluted before testing the pH. To do this, mix a few drops of the substance (or a small pinch of solid) with pure water.</u>

    f. Liquid dish detergent   j. Household ammonia
    g. Dishwasher detergent   k. Liquid laundry bleach
    h. Laundry detergent   l. Baking soda (sodium bicarbonate)
    i. Shampoo or hand soap   m. Drain cleaner (**Caution:** *see note below*)

 CAUTION! Some drain cleaners are extremely caustic. They are designed to dissolve hair and other debris! Be especially cautious about getting any drain cleaner on your skin. Safety glasses are essential. *In case of any skin contact, wash with copious amounts of water and notify the instructor immediately.*

IV. pH of Tap Water, Rain, and Surface Water

1. Measure the pH of tap water. You can try water from the lab, from a drinking fountain, or from elsewhere on campus or where you live.

2. The pH of rain or snow is particularly interesting because of concern about "acid rain," but good measurements are slightly difficult to do. Your instructor may provide samples that have been collected recently or you can set a clean wide-mouth jar outside just before a rain event. Remember that rain is nearly pure water, and this makes the measurements more difficult. The recommended procedure is described in Experiment 19.

3. It is instructive to compare the pH of rain to the pH of river water. If river samples are available in the lab, check the pH of one or more. If they are different from the pH of rain, try to suggest an explanation.

4. If water from a swimming pool is available, check its pH. Is it different from tap water?

V. pH of Water Containing Two Atmospheric Gases: CO$_2$ and SO$_2$

1. Test the pH of water that is saturated with carbon dioxide. One easy way to do this is to use seltzer water that has been left open so that the excess CO$_2$ has escaped and it has gone "flat." Another way is to bubble some CO$_2$ gas (from a pressurized tank) into water. Yet another way is to use the procedure described in Experiment 1. Test the pH of the water containing CO$_2$. Is the solution acidic?

2. Test the pH of water that is saturated with air. Remember that CO$_2$ is only a very small fraction of the gases that make up air. How does your measured pH value compare to the value stated in the textbook?

3. Find out whether sulfur dioxide gas, SO$_2$, produces acidic solutions as claimed in the text.

 CAUTION! Sulfur dioxide is a very toxic gas. In addition, some individuals have a serious allergic reaction to SO$_2$. This test should be done only in a good fume hood and should not be attempted by individuals who are allergic to SO$_2$.

To do this test, it is first necessary to produce SO$_2$ in a Ziplock bag. Fill a plastic pipet with 6 M sulfuric acid and place it in a Ziplock bag. Add about 1–2 grams of sodium sulfite (Na$_2$SO$_3$) to the bag. Close the bag securely, making sure to exclude air, and then squeeze the pipet so that the chemicals will mix and react. With the bag lying on a table, <u>very cautiously</u> open a <u>corner</u> of the bag, insert a dry plastic pipet (with air squeezed out) and fill it with the SO$_2$ gas. <u>Reseal the bag immediately</u>. Slowly bubble this gas into pure water in a test tube or wellplate. Then test the pH of the water. Has the pH changed? Is it acidic or basic? How does it compare to solutions saturated with CO$_2$?   <u>When finished, empty the bag into a waste container in a fume hood.</u>

Clean-up

Leave the pH electrode soaking in water. Dispose of solutions in appropriate waste containers. Clean all glassware, first with plenty of tap water, then finally with pure water. Leave upside down to drain.

Post-lab Questions

1. Explain briefly why pH is a useful way of describing acid and base solutions over a very wide range of concentrations.

2. Were the pH results for HCl and NaOH solutions approximately what you expected? Explain your reasoning. What general rule can you propose for how the pH should change (up or down and by how much) any time the acid concentration increases by a factor of 10? (For example, when HCl concentrations change from 0.0001 M to 0.001 M to 0.01 M, the acid concentrations increase by a factor of 10 each time.)

3. Why was it necessary to dilute substances *f* to *m* before measuring the pH? How do you think the measured pH of the diluted sample compares with the pH of the sample itself?

Name _____   Date _____

Lab Partner _____   Lab Section _____

Data Sheet—Experiment 18

Part II:

Solution	Measured pH	Predicted pH
0.0001 M HCl		
0.001 M HCl		
0.01 M HCl		
0.001 M NaOH		

Part III:

Sample, including source or description	Measured pH

What, if any, conclusions can you draw about the pHs of the different categories of common substances, e.g., (a) pHs of beverages? (b) pHs of soaps and detergents? (c) pHs of substances that normally touch human skin?

Parts IV and V:

Solution or Sample	Source	Measured pH
tap water		
rain		
river or lake		
CO_2 in water		
SO_2 in water		

If Part IV was done: (a) Was the rain acidic or alkaline? Since rain is formed by evaporation and condensation in clouds, why is the pH not that of pure water?   (b) Were tap water and surface water acidic or alkaline? What substances may account for this?

If Part V was done: Were the pHs of CO_2 and SO_2 in water about what you expected? Write chemical equations for the reactions of each gas with water?

Does Acid Rain Fall in My Neighborhood?

INTRODUCTION

Chapter 6 in *Chemistry in Context* examines the problems associated with acidic rain. pH is the primary way of classifying the acidity of rain samples. Studies of the pH of rain and snow are of great importance as part of national and international efforts to monitor the amounts of sulfur and nitrogen oxides in the atmosphere. In this experiment, you will measure the pH of some rain or snow samples. Although studies of rain pH are easy to do, adherence to correct procedures for collecting and handling the samples is essential for reliable measurements. The directions for this experiment focus on the collection and handling of samples. The pH of the samples will be measured as described in Experiment 18 or as outlined by your instructor. Your instructor will supply the exact details of the experiment and instructions for using the type of pH meter available in your laboratory.

Background Information

The pH scale is a convenient way of expressing hydrogen ion concentration in solutions. pH is defined as the negative of the logarithm (to base 10) of the hydrogen ion molarity. In equation form this is

$$pH = -\log M_{H^+}$$

An interesting feature of the pH scale makes use of the fact that **all** water samples contain some hydrogen ions. Even alkaline solutions that have excess hydroxide ion (OH^-) still contain some H^+. The product of the molarities of H^+ and OH^- at 25°C is always 1.0×10^{-14}. In pure water, the hydrogen ion concentration will be 1×10^{-7} molar (as will the hydroxide ion concentration), and the pH will be 7. In acidic solutions, the H^+ ion is in excess, and the concentration will be higher than 1×10^{-7} molar. Therefore, the pH will be lower than 7.

Pure rainwater has a pH of approximately 5.6, due to the presence of carbon dioxide that reacts with water producing carbonic acid that dissociates into hydrogen ions and bicarbonate ions.

$$CO_2(g) + H_2O \rightarrow H_2CO_3 \rightarrow H^+ + HCO_3^-$$

When rain has a pH below 5.6, this means that other acids must be present, and then it is called "acidic rain." Rain can be classified as follows:

Rain classification	pH
Pure rain	5.6
Slightly acidic	5.0 – 5.6
Moderately acidic	4.5 – 5.0
Highly acidic	4.0 – 4.5
Extremely acidic	below 4.0

Overview of the Experiment

1. Prepare sample containers for collecting rain samples.
2. Select sites for the collection containers.
3. Collect rain sample.
4. Prepare the sample for pH measurement.
5. Calibrate the pH meter.
6. Measure the pH of the sample.
7. Analyze the sample data.

Pre-lab Question

The U.S. Environmental Protection Agency prepares maps depicting the pH of rain that falls throughout the United States. These maps can be viewed on the EPA website at:

http://www.epa.gov/castnet/mapnconc.html

Go to the map for the most recent three-year period and use it to predict the pH of the rain you will collect for this experiment. Does the map predict that acid rain falls in your area? Students outside of the U.S. can do a web search for a similar map for their country in order to answer this question.

EXPERIMENTAL PROCEDURE

I. Prepare Sample Containers

Samples should be collected in plastic containers, not glass. It is important that the containers be thoroughly cleaned in advance. The recommended cleaning procedure is to first rinse each container with 6 M hydrochloric acid. This first cleaning is followed by 5 rinses with tap water and 5 rinses with distilled or deionized water. Once cleaned and dried, the containers should be lightly capped or covered to keep them scrupulously clean.

 CAUTION! 6 M hydrochloric acid is a corrosive liquid. Wear gloves and a lab apron along with your safety glasses.

II. Collect Rain Samples

If the bottles have a large opening, they can simply be set out to collect rain. Alternatively, in order to obtain larger samples, it may be desirable to use a large plastic funnel (also cleaned in advance). In this case, the bottle will probably need to be supported in an upright position—for example, by using an empty metal can nailed to a board. The collecting container should be placed in an unobstructed location where it will not be disturbed. Ideally, it should be at least several feet above ground level in order to minimize contamination of the samples by dust and dirt.

A single sample can be collected for an entire rain event. Alternatively, samples can be collected at intervals during the event. For example, for an afternoon thunderstorm, the container might be changed every 15 minutes. For a one-to-three-day rain event, the container might be changed every few hours or twice a day. The pH of rain can change significantly during the course of a rain event.

III. Handling and Storage of Samples

Bottles should be labeled immediately (date, total sampling time, location) and brought to a laboratory for measurement as soon as possible. It is important that each sample be filtered (into another clean bottle!) to remove any dust and debris that might react with the acids in the rain. Ideally, the samples should be filtered immediately after collection. If this is not possible, they should be kept cold until brought to the laboratory and then filtered as soon as possible. If it is not possible to test the samples within a few hours, they should be stored in a refrigerator.

IV. Calibration of the pH Meter

The measurement of pH is normally done with a pH meter. This instrument is convenient and fast to use, but it needs to be calibrated first with solutions of known pH, as described in detail in Experiment 18.

1. Set up the pH meter for calibration by following the instructions provided by your instructor or in an instruction manual.

2. Rinse the electrodes with pure water and immerse them in a standard buffer solution of pH 7.00. Stir for 30 seconds, let stand undisturbed for 30 seconds, and then adjust the meter to read 7.00.

3. Rinse the electrodes thoroughly with distilled or deionized water (a plastic wash bottle is convenient for this). Then immerse the electrodes in a second standard buffer of pH 4.00, stir 30 seconds, let stand 30 seconds, and adjust the appropriate knob to read 4.00.

4. Finally, rinse the electrodes thoroughly and repeatedly with pure water to remove all traces of the buffer solutions. The pH meter is now calibrated and ready to use.

V. Measurement of Rain Samples

The official, recommended procedure from the National Institute of Standards and Technology is as follows: Remove drops of water on the pH electrode by blotting gently. Insert the electrodes in a clean beaker containing 10–20 ml of the rainwater sample. Stir or swirl the solution to ensure homogeneous contact with the electrodes (do not use a magnetic stirrer). Allow the solution to settle for approximately 30 seconds. Record the pH after the reading has stabilized.

Carefully measure the pH values for whatever samples are available. Be sure to rinse the electrodes thoroughly between samples. Record the results on the data sheet.

Clean-up

Leave the pH electrode soaking in water. Dispose of all samples in appropriate waste containers. Clean all glassware, first with tap water, then with distilled or deionized water.

Post-lab Questions

1. Describe the collection site(s) for your sample(s).

2. Collect the class data and classify each of the samples according to the acid-rain scale shown at the beginning of the experiment.

3. If several samples from one rain event are available, how did the pH change during the event?

4. If the samples were collected at different locations, are there any pH differences between locations? If so, how can you explain them?

5. Sometimes, rain samples are found to be not acidic at all and instead are alkaline with a pH greater than 7. What might account for this? If you think you have a possible explanation, what tests could be performed by a chemist to support or refute your idea?

Name _____   Date _____

Lab Partner _____   Lab Section _____

Data Sheet—Experiment 19

Hypothesis

Predict the relative acidity of your rain samples. How do you think acidity will vary based on where and when they were collected?

Measurements

Sample description	pH

Conclusion
Did the pH values of your samples match what you predicted? Explain.

Solubilities: An Investigation
How Do I Design My Own Lab Procedure?

INTRODUCTION

This laboratory exercise is a departure from the usual experiment. It is an attempt to simulate the kind of problem solving that takes place in a scientific laboratory. There are no instructions, procedures, or data sheets. There is simply a problem to solve, which requires reasoning skills and the application of knowledge that has been previously acquired.

A team will typically consist of three or four members, one of whom should be designated as the recorder for the group. This person records ideas, data, experimental procedures, etc. Your group might also want to assign other tasks to members of the group, for example, the "go-fer," the experimentalist, the report writer, the team leader. (Your investigation will be more efficient if you take a few minutes at the beginning to be sure each person has a clear role to play.) You and your team members must come up with a solution to the problem and then be prepared to compare your method and your results with those of the other teams in the class.

THE PROBLEM

Determine the solubilities of the following compounds in water at room temperature and rank them in order from least soluble to most soluble: calcium sulfate ($CaSO_4$), potassium aluminum sulfate [$KAl(SO_4)_2$], potassium nitrate (KNO_3), ammonium nitrate (NH_4NO_3), copper nitrate [$Cu(NO_3)_2$] and sodium chloride (NaCl). All of these compounds, except the copper salt, are likely to be found in drinking water supplies since they occur naturally in rocks and soil or are used as components in fertilizers. Copper nitrate is included because copper salts are often used to reduce algae in ponds and swimming pools.

Solubility is defined as the maximum mass of solid that can dissolve in a given volume of water. It is usually reported as grams of solid per mL of water or grams of solid per 100 mL of water. Think carefully about how the procedure you design will help you measure this information.

In the interest of cost and convenience, you may use no more than 2 grams of each compound. Develop a procedure for carrying out this task and record it and any observations you make while solving this problem. Your grade will not depend on getting a "correct answer," but rather on the quality of the procedure you develop and the care with which you carry out your investigation.

Materials and Equipment

A variety of materials and equipment will be available that you can use to solve this problem. At a minimum, your "lab" should have graduated cylinders, burets, beakers, flasks, plasticware, stirrers, test tubes, plastic pipets, and a balance.

Reporting Your Results

When everyone is finished, your class will assemble to hear a report from each student team about the method used and their results. Therefore, it is important for your team to keep a complete record of everything you do and the numerical data you obtain. In addition to an oral report to the class, your group should prepare a brief (one- to two-page) written report of your investigation. Your instructor will specify what should be included in the report.

Measurement of Radon in Air
Am I Breathing Radioactive Particles?

INTRODUCTION

Radon is a radioactive gas that is a potentially serious indoor health hazard in some geographic areas. It is discussed in Sections 1.13 and 7.6 of *Chemistry in Context*. In this experiment, you will use a simple radon detector to measure the radon concentration in a location of your own choosing. A sampling period of at least 4 weeks is necessary in order to obtain reliable data.

Background Information

The atomic number of radon is 86, which places it in group 8A, at the far right-hand side of the periodic table. Elements in group 8A are all gases that are chemically unreactive and generally do not form molecules. Radon is an intermediate product in the radioactive decay of uranium-238. (The U-238 decay series is shown in Figure 7.12 in *Chemistry in Context*.) Uranium is widely distributed on the surface of our planet. For example, granite rocks contain about 4 ppm of uranium. Radon-222 (the longest lived isotope of radon) is an alpha-emitter, with a half-life of 3.8 days. Once produced from uranium, radon is drawn upwards through fissures in the rocks and soils. (Because radon is chemically inert, it passes through soil without reacting.) Radon can enter homes through cracks or other holes in the foundation and is typically more serious in houses with basements. The amount of radon found in homes varies from one part of the country to another. National and state maps of predicted radon concentrations in homes can be found at the following EPA website: http://www.epa.gov/radon/zonemap.html.

In this experiment, you will use a small plastic disk to detect radon. The disk consists of a high-clarity polymer often used in eyeglasses that is known in the trade as "CR-39." The full chemical name for CR-39 is poly[ethylene glycol bis(allyl carbonate)]. As explained in Chapter 9 of *Chemistry in Context*, polymers are large molecules built from smaller units called monomers.

Alpha particles (see Table 7.1 in the text) are a type of ionizing radiation. When they penetrate CR-39, they cause damage in the plastic, probably due to disruption of the polymer chain along the path of penetration. Although the damage is not visible to the eye, it can be revealed by treating the disk with sodium hydroxide, NaOH. Sodium hydroxide etches the sample preferentially in the damaged regions. The etched regions show up as "tracks" when viewed under a microscope. The number of tracks in a given area of the plastic disk can be used to estimate the radon level at the sampled location. Although beta and gamma radiation also penetrate the plastic, these rays mostly pass through without causing damage.

Overview of the Experiment

1. Place a detector disk in an undisturbed location for at least 4 weeks.
2. Etch the disk to produce visible alpha tracks.
3. Count the tracks under a microscope.
4. Calculate the concentration of radon in air by comparison with a control disk.

Pre-lab Question

Alpha particles are nuclei of which element?

EXPERIMENTAL PROCEDURE

I. Air Sampling for Radon

Start the air sampling early in the semester as directed by your instructor. You need to use as long a sampling period as possible, preferably for *at least* a month. Select a location where the detector can remain undisturbed and where you can retrieve the detector prior to the scheduled lab time for this study. (*Plan this carefully, perhaps adding a label to explain what the item is. Janitors or other persons may unknowingly remove the disk when cleaning up an area!*) The class results will be more interesting if students select many different locations, both off-campus and on-campus, inside buildings and outdoors, different floor levels in buildings, different kinds of ventilation or air circulation systems. One possibility is to use your own home, either taking the detector device there yourself or mailing it to a relative with instructions. In some parts of the country, basements of houses are particularly prone to high levels of radon.

You will be given the following materials:

1. A small piece of CR-39.
2. A small plastic cup, preferably with lid.
3. A small rectangle of cardstock (size depends on the cup being used).
4. Piece of "Scotch" tape.
5. Small piece of thin, single-layer toilet or tissue paper (to protect the CR-39 from dust).

Proceed as follows, once arrived at the sampling location:

1. If the cup has a lid, cut a large circular hole in the lid.

2. Remove the plastic film from the side of the CR-39 that is marked.

3. Make a loop of tape with the sticky side out. Stick the side of the CR-39 that still has a plastic film coating onto the tape. Then stick the CR-39 to the middle of the piece of cardstock.

4. Put the cardstock plus CR-39 into the plastic cup, with the CR-39 pointing up and the card bent in an inverted U so that the CR-39 is near the top of the cup. Cover the plastic-cup opening with the toilet or tissue paper. Then snap the lid onto the cup. (If the cup does not have a snap-on lid, use a rubber band to hold the toilet paper in place.)

5. Record the start date and time.

6. At the end of the air-sampling period, record the ending date and time. Store the card with the attached disk in a labeled envelope until time for the laboratory work.

Note: It is important that you record accurately the location for the detector disk (describing it in detail) and the start and stop times (dates and time of day).

II. Etching the Disk

Remember to bring your exposed disk to your laboratory class!

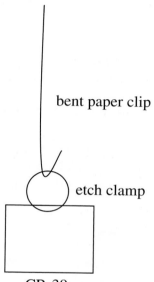

1. Remove the disk from the card to which it has been attached. You already peeled off one of the polyethylene films from the disk. Now peel off the other (from the back of the disk). Gently slip the ring of an "etch clamp" over the top of the disk. (*Caution:The disk is fragile and can break easily.*) Then fasten the ring onto a paper clip that has been bent so there is a hook at each end (see *Figure 21.1*).

2. Hook the disk inside a test tube and add enough 6 M sodium hydroxide solution (NaOH) to cover the disk in the tube. *CAUTION: The coating on the paper clip will react with NaOH; therefore, try not to submerge any more of the paper clip than is necessary.*

 CAUTION! Hot 6 M NaOH is extremely corrosive to skin and to clothing. *Wear goggles at all times. Rubber gloves are strongly recommended.*

Figure 21.1
Experimental set-up for etching.

3. Start a beaker of water heating on a hot plate and bring to a boil. Heat the test tube containing the disk and NaOH solution in this boiling water bath for 40 minutes. At the end of 40 minutes, remove the disk and rinse thoroughly with lots of tap water.

4. **Clean-up:** When finished, do NOT pour the NaOH solution down the drain. Pour it into an approved container provided by the instructor.

III. Counting the Alpha Tracks

This will be done with a microscope. Your instructor or teaching assistant will show you how to operate the particular microscope that you will be using.

1. Determine the area (in square centimeters) of the field of view of your microscope at low power (10x) by looking through the microscope at a ruler. (Consult your instructor if you are unsure about how to calculate this.) To do this, count the number of millimeter divisions on the ruler that you can see across the widest portion of the field of view and record this. This is the diameter in millimeters. Divide by 10 to find the diameter in centimeters. The radius, r, is half of this, and the area, A, can be calculated by the formula $A = \pi r^2$.

2. <u>General instructions for counting alpha tracks</u>. Place an etched disk on a microscope slide under the microscope. Then adjust the focus until you are sure you are focused on the etched disk. Since the glass slide and disk are at different heights, they will come into focus at different times as you adjust the focus knob on the microscope. As a result, it is easy to mistake the microscope slide for the disk. To prevent this from occurring, move the microscope slide so that the edge of the disk is in view. Then adjust the focus until the disk comes into sharp focus. Once you have focused on the disk, you can move the slide around to observe the tracks in different areas.

3. Practice looking at alpha tracks on a previously exposed and etched disk. This may be a disk donated by a student in a previous class, or it may be one prepared by your instructor using a radioactive source (in which case, your own disk will likely have fewer tracks).

    Look at the two pictures of alpha tracks at different magnifications (*Figure 21.2*). The right-hand photo (*Figure 21.2B*) is more nearly what you will see. The various shapes of the tracks depend on how the alpha particles entered the solid (see *Figure 21.2A*). The circular-shaped tracks are due to alpha particles that entered straight (perpendicular to the disk), and the more teardrop-shaped tracks are due to alpha particles that entered at an angle.

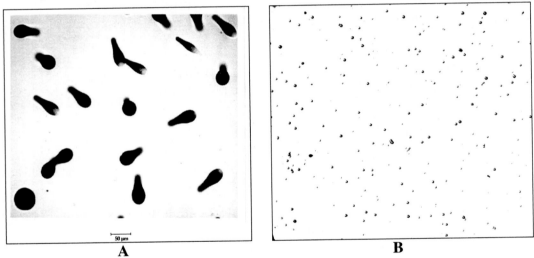

A   B

Figure 21.2 Appearance of alpha tracks on an etched CR-39 disk. (A) At high magnification, showing different shapes of tracks. (B) 20X magnification, sample exposed to 9 pCi/L for five days.

4. Place your own disk on a microscope slide; look through the microscope and adjust the focus until you are sure you are focused on the etched disk. Again, be sure not to mistake the microscope slide for the disk. Once you have focused on the disk, you can move the slide around to observe the tracks in different areas.

5. Count the number of tracks *in 10 different fields of view* and record them on the data sheet.

IV. Calculations

1. Determine the average number of tracks in a field of view for your disk.

2. Then determine the average number of tracks per square centimeter per day of exposure (including any fraction of a day). See your instructor if you need help with this calculation.

3. Control plastic disks that were sent to a radon facility in which the radon level was known to be 370 picocuries per liter were found to exhibit 2,370 tracks/cm²/day by this etching and counting technique. Given this relationship between picocuries per liter and tracks per square centimeter per day, calculate the radon level in picocuries per liter of air for *your* sample and enter this on the data sheet. How does your sample compare with the 4-pCi/L guidelines set by the Environmental Protection Agency?

Note: Concentrations greater than 4 pCi/L indicate the need for a follow-up test. You can do another one yourself or use an EPA-approved radon monitoring service. If a concentration greater than 4 pCi/L is found again, you should consult with experts for more detailed analysis and mitigation.

Post-lab Questions

1. For many types of chemical measurement, you need to subtract out a naturally occurring background level of the chemical being measured. In this case, you don't expect to be able to measure any background radiation from alpha particles, so we don't have to subtract out a background. Why don't we expect alpha particle background radiation?

2. You are comparing your measurements to those in picocuries. What does pico mean? What does a curie measure? See your textbook if needed.

3. Answer each of the following three questions, either individually or in a team as assigned by your instructor.

    a. Consider a beachside home in Mayagüez, Puerto Rico, a university town on the west coast of this volcanic Caribbean island. The home is built on a concrete slab and has screen windows. Is radon likely to be a problem for a home-owner in Mayagüez? Why or why not?

    b. Consider a ten-story office building in chilly Duluth, Minnesota. The building is well insulated and sealed against the winter arctic blasts. Is radon likely to be a problem for the occupants of this building? Why or why not?

    c. Consider a two-story farmhouse in rural Nebraska. The basement has a dirt floor with underlying limestone rock. Is radon likely to be a problem for the farmers who have lived there for generations? Why or why not?

4. Here is the health advisory from the surgeon general:

    "Indoor radon gas is a national health problem. Radon causes thousands of deaths each year. Millions of homes have elevated radon levels. Homes should be tested for radon. When elevated levels are confirmed, the problem should be corrected."

    Consult your textbook and list 3 things that can be done to mitigate high radon levels. Explain briefly how each of the actions you list will help solve the problem.

5. Visit www.epa.gov/radon/zonemap.html. Find your home and college on the maps. What are the expected levels of radon in your area? Should your family be concerned about radon?

Name _____   Date _____

Lab Partner _____   Lab Section _____

Data Sheet—Experiment 21

Part I: Air Sampling

Location of detector: _____

Detector disk put in place   Date_____ Time _____

Detector disk removed   Date _____ Time _____

Total days of air sampling (to nearest day)_____

Hypothesis

Look at the maps on the EPA website cited in the introduction. Does the map predict that radon will be found in the area where you placed your sampling disk?

Part III: Calculation of field of view in microscope

# of millimeters visible on the ruler	Diameter of view (cm)	Area of view (cm^2)

Observed number of tracks (10 different fields of view)

Part IV: Calculations

Average number of tracks per field of view	
Number of tracks per cm^2	
Number of days sampled	
Number of tracks per cm^2 per day	
Radon concentration, pCi/L	

Conclusion

How does this value compare to the guideline of 4 pCi/L set by the Environmental Protection Agency? Is further action warranted? Is your calculated radon concentration consistent with your hypothesis?

Can We Get Electricity from Chemical Reactions?

INTRODUCTION

This experiment is designed to accompany Chapter 8 in *Chemistry in Context*. You will construct several simple electrochemical cells in which chemical reactions produce electricity similar to the galvanic cell shown in *Chemistry in Context* Your Turn 8.4. You will also investigate the opposite process in which electricity is used to produce chemical reactions.

Background Information

Any spontaneous chemical reaction involving transfer of electrons from one atom or molecule to another atom or molecule can be used as the basis for an electrochemical cell. Consider for example the reaction between zinc metal and copper ions in water solution:

$$Zn(s) + Cu^{2+}(aq) \rightarrow Zn^{2+}(aq) + Cu(s)$$

If the reaction is allowed to occur in a single container, for instance, by adding zinc metal, $Zn(s)$, to a water solution containing copper ions, $Cu^{2+}(aq)$, the blue color of the copper ions disappears, the zinc gets used up, and the reddish color of metallic copper appears. Electron transfer has occurred, with each zinc atom losing two electrons (to form the +2 ions) and each copper ion gaining two electrons. As explained in the text, we can break this up into two half reactions:

$$Zn(s) \rightarrow Zn^{2+}(aq) + 2\ e^-$$

$$Cu^{2+}(aq) + 2\ e^- \rightarrow Cu(s)$$

In order to get useful electric energy from this reaction, the two half reactions have to occur in two separate locations but connected in such a way that the electrons will flow through an external wire. The wire can be connected to something (such as a motor, lightbulb, or meter) to show that an electrical current is produced. Such a device is correctly called a **galvanic cell**, though the nonscientific and slightly incorrect name, *battery*, is often used. You will investigate how to assemble several working galvanic cells, and you will use a voltmeter to measure the voltages produced.

Rather than allowing a chemical reaction to produce electric energy (in the form of a flowing electron current), the reverse is also possible. In an **electrolytic cell**, electric energy from some source (e.g., a commercial "battery") is used to force a chemical reaction to go backwards, i.e., in the nonspontaneous direction. See *Chemistry in Context* Section 8.6 and *Figure 8.15* for an electrolytic cell used to split water.

Both types of cells have many practical applications. Everyone is familiar with electrochemical cells ("batteries") ranging in size from the tiny cells in hearing aids or watches to large automobile or truck batteries. Electrolysis cells are widely used in chemical manufacturing to produce chemicals that would be otherwise difficult to make.

Overview of the Experiment

1. Assemble 3 galvanic cells and measure their voltages.
2. Construct and test a "citrus cell."
3. Assemble a simple electrolysis unit.
4. Carry out the electrolysis of water.
5. Carry out the electrolysis of potassium iodide solution.
6. Interpret the two electrolysis reactions.

Pre-lab Questions

Locate the three metals used in today's experiment on the periodic table. What are their atomic numbers? Based on what you know about periodicity, how do you think these metals will react? Which other metals might be good for making galvanic cells?

EXPERIMENTAL PROCEDURE

Note: This experiment has two parts that can be done independently and in either order. To facilitate efficient use of equipment, your instructor may assign half of the class to start with Part I (galvanic cells) and the other half to start with Part II (electrolytic cells).

I. Galvanic Cells (electricity from chemical reactions)

You will set up three different galvanic cells:

1. $Zn \rightarrow Zn^{2+} + 2\ e^-$ with $2\ Ag^+ + 2\ e^- \rightarrow 2\ Ag$

2. $Zn \rightarrow Zn^{2+} + 2\ e^-$ with $Cu^{2+} + 2\ e^- \rightarrow Cu$

3. $Cu \rightarrow Cu^{2+} + 2\ e^-$ with $2\ Ag^+ + 2\ e^- \rightarrow 2\ Ag$

In order not to waste chemicals, these will set this up at a small scale using a piece of filter paper placed on a watch glass.

1. Obtain a watch glass and a piece of filter paper of the same size. On the filter paper, draw three small circles, labeling them Ag^+, Cu^{2+}, and Zn^{2+} as shown in *Figure 22.1*. Trace a trail from each circle to the center of the paper. Use scissors to cut out wedges of paper between the circles. Place the filter paper on the watch glass.

2. Obtain small pieces of each of the three metals: Ag, Cu, and Zn. Using sandpaper, sand both sides of each piece of metal until it is clean and shiny.

3. Place 3 drops each of 0.1 M solutions containing Cu^{2+}, Zn^{2+}, and Ag^+ on the corresponding circles. Then place each piece of metal on the spot of its corresponding ion. *The top sides of the metals should be kept dry.*

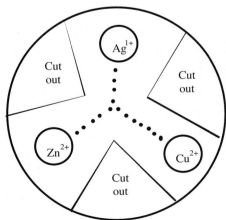

Figure 22.1
Galvanic cell assembly

NOTE: Silver nitrate will produce a black stain on skin, although it is not harmful. Try to avoid getting this solution on your skin.

4. Now attempt to measure the voltage for the zinc and copper combination as follows. Obtain a voltmeter and learn how it operates. Turn it on and select a range of about 1 or 2 volts. (Some meters automatically select the appropriate range.) Check to be sure that the meter reads zero volts when the two probes are touched together. Then touch the two metal probes from the meter to the pieces of copper and zinc and observe the voltage.

5. It is probable that you observed zero volts (or very close to zero), and you should convince yourself that there is a missing link that prevents a complete cycle or circuit. To complete the circuit, fill a small plastic pipet with 1 molar KNO_3 solution and carefully dribble a small amount of solution along the lines connecting the circles to the middle of the paper. Be sure there is a continuous trail of KNO_3 connecting the circles.

6. Test the voltage again by touching the probes to the Cu and Zn metals. If a positive voltage is displayed, wait 5 seconds and record the reading. If a negative voltage is displayed, reverse the leads, wait 5 seconds, and record the reading. Also record which metal the red probe touches to give a positive voltage.

7. In a similar fashion, measure the voltage for the other two possible combinations: copper + silver and zinc + silver. Record the voltages and which metal the red probe touches to give a positive voltage.

8. The Citrus Cell. The metal ions that were supplied by solution in the last part of this experiment are available from many natural sources, e.g., very hard water or fruits and vegetables. To illustrate this, you can make a "battery" from a piece of citrus fruit. Choose two of the three metals (Zn, Cu, Ag). Push the metal strips through the fruit rind so that a part of each is inside the juicy portion of the fruit. The two pieces of metal must not touch but should be in the same section of the fruit. Record the voltage and which metals you used.

Further Explorations

This study could be expanded to include other metals. Among the possibilities are iron, nickel, tin, and magnesium. Some metals such as aluminum form a very tough coating that makes this kind of measurement impractical. Certain other metals such as sodium and potassium are so reactive that they will immediately react with the water!   Write a brief description of the exploration you plan, have your instructor check it for safety and feasibility, and, if approved, carry it out.

Clean-up

When done, rinse the small metal pieces with pure water and blot dry with a paper towel. Dispose of the used filter paper in an appropriate waste receptacle. Rinse off the watch glass with pure water. Return metals and watch glass to the supply location.

II. Electrolytic Cells (chemical reactions from electricity)

Assemble a simple electrolysis unit using two 9-volt batteries, as shown in *Figure 22.2*. (A single 9-volt battery can be used, but two batteries connected in "parallel" are preferable and will give more dramatic results.)

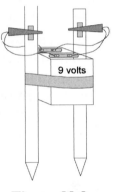

1. Tape two 9-volt batteries together.

2. Obtain two snap-on battery connectors and two alligator clips, one red and one black. Twist the ends of the two red wires together. Thread these combined wires through the hole in the handle of the red alligator clip. Bend the wire back on itself and twist to make it snug. Repeat this process for the black wires.

Figure 22.2 Electrolysis unit

3. Obtain two graphite rods. (Ordinary pencils with a good point work well for this but must have a place to attach an alligator clip, either by sharpening the top end or by cutting away the wood near the middle.)

4. Tape the rods or pencils to either side of the batteries. Snap the battery connectors onto the batteries and clip one alligator clip to each graphite rod.

A. Electrolysis of Water. Use your electrolysis unit to split water (H_2O) into hydrogen (H_2) and oxygen (O_2).

1. Obtain a Petri dish. Fill it about half full with distilled or deionized water. Add 2 drops of phenolphthalein indicator and swirl gently to mix. This indicator is colorless in neutral or acidic solutions (excess H^+) and pink in basic solutions (excess OH^-). (Your instructor may provide a different indicator and, if so, will tell you the color changes.)

2. Immerse the tips of the graphite rods and observe carefully. Do you see any evidence of a chemical reaction?

3. Add about one-half teaspoon (about 1 gram) of solid potassium nitrate, KNO_3. Jiggle the dish gently to aid in dissolving the KNO_3 crystals. The job of the KNO_3 is to be the electrolyte, or to help the solution conduct electricity by providing K^+ and NO_3^- ions, <u>without</u> directly participating in any reactions.

4. Again immerse the graphite rods in the solution. Observe carefully for a couple of minutes. You should see evidence of chemical change. (Set the dish on a light-colored background to make the observations clearer.) Record what is happening at the electrode attached to the red wires and what is happening at the electrode attached to the black wires. Which one produces more gas bubbles? Which causes the indicator to change color?

5. Dispose of the solution in an appropriate waste container.

B. Electrolysis of a Water Solution of Potassium Iodide, KI(aq)

1. Using the same dish from Part A, add about one-half teaspoon (about 1 gram) of solid potassium iodide, KI, to the dish.

2. Fill your Petri dish at least half full with tap water. Jiggle the dish gently to aid in dissolving the KI crystals. The KI does two jobs here. It serves as the electrolyte, helping the water conduct electricity, but it also will be a reactant.

3. Immerse the graphite electrodes in the solution. Record your observations, noting what is happening at or near each of the two electrodes.

4. While the electrolysis is running, add 2 drops of phenolphthalein indicator directly next to each electrode. As noted previously, this indicator is colorless in neutral or acidic solutions (excess H^+) and pink in basic solutions (excess OH^-). Record any color changes you see and note whether this was at the red or black wire electrode.

Clean-up
1. Dump the Petri dish contents in the designated waste container.
2. Return the electrolysis unit or disassemble it if instructed to do so.

Post-lab Questions

1. Explain why the KNO_3 solution was needed for your galvanic cells to function.

2. Voltages can be increased by hooking together several cells in a series. How many of the Cu/Zn cells would you need to combine to get the same 1.5 volts you get from a purchased dry cell?

3. What voltage did your citrus cell give? How many of these would you need to get a voltage equivalent to a dry cell?

4.   Imagine that you are marooned on an isolated island with a few other people. After doing this experiment, what could you offer to the group that would be useful for supplying energy? Suggest something specific you could do with a few pieces of metal and food scraps or some nearby salt deposits.

5. The two half-reactions for the electrolysis of water are

$$4 H_2O(l) + 4 e^- \rightarrow 2 H_2(g) + 4 OH^-(aq)$$ (2 moles of gas and a basic solution formed)
and
$$2 H_2O(l) \rightarrow O_2(g) + 4 e^- + 4 H^+(aq)$$ (1 mole of gas and an acidic solution formed)

Based upon your observations, which of the two half-reactions appears to be occurring at the black-wire lead? Explain your reasoning. Which gas formed at which color electrode? Explain how you decided.

6. In light of your answer to #5 and the equations for the half-reactions, which color electrode was supplying electrons from the batteries? What was the other electrode doing? Explain.

7. In Part II-B, during the electrolysis of the KI solution, either H_2O or KI could react at each of the electrodes. Thus, although only one reaction will actually occur at each electrode, we must consider two possible reactions for the red electrode and two possible reactions for the black electrode. The half-reactions that could occur are the following:

at the electrode where electrons are consumed:

Either: $4 H_2O(l) + 4 e^- \rightarrow 2 H_2(g) + 4 OH^-(aq)$ (gas and basic solution formed)

Or: $K^+(aq) + 1 e^- \rightarrow K(s)$ (metallic potassium formed)

at the electrode where electrodes are released:

Either: $2 H_2O(l) \rightarrow O_2(g) + 4 e^- + 4 H^+(aq)$ (gas and acidic solution formed)

Or: $2 I^-(aq) \rightarrow I_2(aq) + 2 e^-$ (yellow-brown elemental I_2 formed)

a. From your observations, what formed at the black-wire lead? Explain your reasoning.
b. From your observations, what formed at the red-wire lead? Explain your reasoning.
c. Now compare your electrolysis of water to that of the KI solution. Which product from the water electrolysis was also produced in the KI electrolysis?
d. Which product from the water electrolysis was NOT produced in the KI electrolysis? What was formed instead? Electrolysis will always produce the products that are easiest to make from the starting materials. What can you conclude about how easy it is to remove electrons from iodine in KI versus how easy it is to remove electrons from oxygen in H_2O?

Name _____   Date _____

Lab Partner _____   Lab Section _____

Data Sheet—Experiment 22

Part I: Electrochemical (voltaic) Cells

Zn/Cu cell voltage = _____   Red probe on which metal? _____

Zn/Ag cell voltage = _____   Red probe on which metal? _____

Cu/Ag cell voltage = _____   Red probe on which metal? _____

Sketch a diagram of your Cu/Zn cell showing the metals and solutions involved as well as the voltmeter. Show the direction of electron flow out of one metal and through the voltmeter to the other metal. (Recall that the red-wire probe was on the metal that consumes electrons.) Show the rest of the path that negative charge must follow to complete the circuit.

Rank the three cells above in order by the voltages they generated. Include in your rankings a standard Zn/MnO_2 dry cell (1.5 volts), similar to what is used in flashlights.

1. _____

2. _____

3. _____

4. _____

Which metals will you choose to make your citrus cell? Why?

Citrus cell voltage = _____

Part II: Electrolytic Cells

A. Electrolysis of water:

   Observations at the black-wire electrode

   Observations at the red-wire electrode

What two gases are the final products from the electrolysis of water? Write a balanced equation for the electrolysis of water. (Water is the only reactant, and the gases are the products.)

B. Electrolysis of potassium iodide solution:

   Observations at the black-wire electrode

   Observations at the red-wire electrode

How Do Polymer Properties Connect to Structure? Polymer Synthesis and Properties

INTRODUCTION

Polymers are long molecular chains composed of smaller repeating units. Natural polymers such as proteins, starch and cellulose are important to all living systems. Synthetic polymers, such as polystyrene and nylon, also are important and have both industrial and household uses. Although the polymers that we will study are composed primarily of carbon, their properties vary widely. We can make polymers that are opaque or transparent, rigid or flexible, weak or strong, sticky or smooth.

Today, both producers and consumers are becoming more concerned about the toxicity of polymers in household products, and about the ultimate fate of polymers in recycling centers or in the landfill. Chapter 9 of your text, *Chemistry in Context*, provides more information. In this experiment, you will examine the properties of several polymers, including some that you will synthesize. You also will relate the properties of polymers to their molecular structures.

Background Information

The properties of a polymer depend not only upon the nature of its monomer, or repeat unit, but also on the arrangement of the polymer chains relative to each other. In most polymers, the molecules are arranged randomly, as shown in *Figure 23.1*. The strength of such polymers will be equal in all directions. In other polymers, however, the molecular chains are arranged in parallel to each other, as shown in *Figure 23.2*.

Figure 23.1 Random polymer chains

Figure 23.2 Parallel polymer chains

The strength of a polymer with parallel chains varies with direction. Pulling lengthwise on a piece of this polymer meets with resistance from the covalent bonds within a chain. Pulling crosswise, however, has less resistance and simply moves one chain further away from another.

In some polymers, one molecular chain is only weakly attracted to its neighboring chain. This results in polymers that are more flexible or that have lower melting points. Stronger attractions between chains result in solid polymers that are more rigid or crystalline and liquid polymers that are more viscous (syrupy). In crosslinked polymers, the chains are covalently linked together. This results in non-crystalline polymers that have high viscosity and mechanical strength. In

Part I, you will study the properties of three polymers with different structures to determine how the structure affects the strength of the polymer.

In Part II, you will prepare nylon, a very strong fiber described in Section 9.6 of *Chemistry in Context*. You will do a condensation reaction of a diacyl chloride with a diamine.

| diacyl chloride | diamine | Nylon-66 |

In Part III, you will prepare polyurethane foam. This polymer has multiple growth points on each strand, and thus, the polymer is able to branch and form crosslinked chains. A gas produced by a side reaction causes the polyurethane to foam.

In Parts IV and V, you will add sodium borate to polyvinyl alcohol, creating cross-links, or bonds, between adjacent polymer strands, as shown in the equation below. More abundant and stronger cross-link bonds give polymers that are more rigid or more viscous.

Throughout this experiment, you will be asked to observe and record properties of polymers. The properties that are noteworthy will vary, but examples you should consider include

- color
- physical state (liquid or solid?)
- viscosity (how syrupy?)
- brittleness (would it break if hit hard?)
- elasticity (will it stretch and snap back?)
- adhesion (sticky or smooth?)
- tensile strength (how hard can you pull before it breaks?)

Overview of the Experiment

1. Observe properties of Teflon® tape, HDPE, and Mater-Bi before and after stretching.
2. Observe degradation of Teflon, HDPE and Mater-Bi with acid.
3. Prepare and observe properties of nylon.
4. Prepare and observe properties of polyurethane foam.
5. Prepare and observe properties of polyvinyl alcohol gel.
6. Measure effect of different amounts of sodium borate on polyvinyl alcohol gel viscosity.

Pre-lab Questions

Draw the structure of one repeat unit of each polymer you will use or make in this experiment. Identify the elements present. Which ones just contain carbon and hydrogen? Which other elements are found in each polymer?

EXPERIMENTAL PROCEDURE

SAFETY NOTES

- Avoid getting the solutions used to prepare nylon or polyurethane on your skin. If you do get these on your skin, wash them off with large amounts of water and notify your instructor
- The polymers you make may be handled directly, but you should thoroughly wash your hands afterward.

I. Examination of Polymer Films

1. **Teflon**. Obtain a piece of polytetrafluoroethylene (PTFE, or Teflon) tape 5 cm in length. Teflon is commonly used as non-stick coating on cookware, though a molecule used in its processing, Observe the properties of this polymer. Hint: revisit the Background section for suggested properties to consider. Test the strength of the polymer by gently pulling first along its length and then pulling side-to-side. Record your observations. **Note:** Teflon is commonly used as a non-stick coating on cookware. A compound used in its processing, perfluorooctanoic acid (PFOA), is currently the source of a public controversy.[1]

2. **HDPE**. Obtain a piece of high-density polyethylene film (roughly 5 cm x 10 cm) that has been cut from a grocery-store bag. HDPE is a linear form of polyethylene (in contrast to low-density polyethylene, LDPE, which is branched).

    a. Observe and record the properties of this polymer. Test the strength of the polymer by gently pulling first lengthwise and then side-to-side. Record your observations.

    b. Now grasp the film by its shorter sides. Pull firmly and slowly until the middle of the strip has stretched significantly. Observe and record the properties of this stretched section. Compare the width of the stretched section to that of the unstretched end portions. Test the strength of the stretched portion both lengthwise and crosswise. Record your observations.

3. **Mater-Bi**. Obtain a piece of Mater-Bi film that has been cut from a biodegradable BioBag. Mater-Bi is a natural and renewable polymer formed by crosslinking amylose, a linear form of starch, to varying degrees to obtain polymers with different properties. The piece you have been given is from a bag used to collect food waste for community composting projects. Test the properties of the Mater-Bi as you have the previous two polymers and record your observations.

---

[1] New Jersey Department of Environmental Protection, *Determination of Perfluorooctanoic Acid (PFOA) in Aqueous Samples*, Final Report, January 2007. http://www.state.nj.us/dep/watersupply/final_pfoa_report.pdf

4. **Degradation of polymers**. Place a small piece of each of the three polymers into separate wells of a wellplate. Cover each sample with enough 3 M H_2SO_4 (sulfuric acid) to cover the polymer. Let the polymers sit while you do the rest of the experiment, or for at least 45 minutes. After this time, examine the properties of the polymers again. Have any of the polymers been affected by reaction with sulfuric acid? How have the properties changed? Which of the polymers seem to be resistant to reaction with sulfuric acid?

II. Synthesis of Nylon

1. Obtain 10 mL of nylon solution A (adipoyl chloride in hexane) in a 50-mL beaker. In a second 50-mL beaker, obtain 10 mL of nylon solution B (hexamethylenediamine in water).

2. Tip the beaker containing solution B at a slight angle and then *slowly* add solution A by gently pouring it down the side of the beaker. Solution A should form a separate layer on top of solution B. Do NOT stir or mix. A film of nylon will form between the layers.

 STOP! Do not touch the nylon with your bare hands until it has been thoroughly rinsed with water. Hydrochloric acid is formed as a reaction by-product and can burn.

3. Use tweezers to grasp the film. Gently pull it up and out of the beaker. Wrap the film around a test tube and then rotate the test tube to gradually pull the rest of the film from the beaker. If you are careful, you can get one long continuous strand. If the strand breaks, simply grab it again with tweezers and wrap it around the test tube again. Continue pulling nylon film until the solutions are used up.

4. Rinse your nylon thoroughly under a gentle stream of tap water. Slide the nylon off of the test tube. Observe and record the properties of the wet nylon.

5. Spread some of your nylon strands out on a paper towel to dry. Continue with other parts of the experiment but return to observe and record the properties of the nylon after it dries.

6. When done, dispose of the solid nylon and the excess solutions as directed by your instructor.

III. Synthesis of Polyurethane Foam

1. Obtain 4 mL of polyurethane solution A in a 3-oz. paper cup. Be patient as you pour so that you get as much of the viscous liquid into the cup as possible. If desired, add 2 drops of food coloring. Mix with a wooden splint or disposable applicator stick.

2. Add 4 mL of polyurethane solution B to the paper cup. Use the wooden splint to <u>thoroughly</u> mix the liquids. Once the liquids are well mixed, remove the mixing stick and allow the cup to sit undisturbed.

3. The reaction may take a minute or two to begin. Once it does, record your observations. You may gently touch the outside of the cup as the reaction proceeds but do not touch the polymer itself yet.

4. After about 5 minutes, the reaction will be complete, and you may touch the polymer, which should no longer be sticky. Observe and record the properties of the polyurethane foam.

5. When done, dispose of the polyurethane foam as directed by your instructor.

IV. Synthesis of Polyvinyl Alcohol Gel

1. Obtain 50 mL of 4% polyvinyl alcohol solution in a 100-mL beaker. (Pure polyvinyl alcohol is a white solid, but it has been dissolved in water here.) Observe and record the properties of this polymer solution.

2. Obtain 5 mL of 4% sodium borate solution. Add this to the polyvinyl alcohol with stirring. Continue stirring until the mixture becomes homogeneous. This may take several minutes. You may wish to use a metal spatula rather than a glass rod to stir as the mixture becomes very thick.

3. Remove the polymer from its beaker and examine it. You can touch the polymer with your hands, just be sure to wash your hands afterwards. Try forming a ball with the polymer. Does it bounce? Does it remain round? Does it stretch? Observe and record the properties of the polyvinyl alcohol gel.

4. When done, dispose of the polyvinyl alcohol gel as instructed.

V. Viscosity of Polyvinyl Alcohol Gels

In this portion of the experiment, you will determine how the amount of sodium borate solution added to polyvinyl alcohol affects the viscosity of the polymer. Your instructor will assign you a specific amount of sodium borate to add. Work carefully since your results will be combined with those of the rest of the class.

1. Obtain 50 mL of 4% polyvinyl alcohol solution in a 100-mL beaker as before.

2. Obtain your assigned volume of 4% sodium borate solution (from 0 to 5 mL). Add enough water to bring this solution to 5 mL. Mix.

3. Note the time and then add the sodium borate solution to the polyvinyl alcohol solution. Mix thoroughly. It is important to get this gel as homogeneous as possible. A metal spatula may be useful to break up solid lumps. Periodically mix and break up lumps for 5 minutes.

4. After mixing for 5 minutes, transfer your gel into a 50-mL beaker. Use a marker and ruler to place two lines 30 mm apart on the side of the beaker. The top line should be somewhat below the surface of the gel and the bottom line should be somewhat above the bottom of the beaker.

5. Hold a steel ball above your gel so that it just touches the surface. Release the ball and record the time in seconds that it takes the ball to move from the top line to the bottom line.

6. As directed by your instructor, post your data (volume of sodium borate and time for the ball to drop) for use by the rest of the class.

7. Dispose of your gel mixture as directed by your instructor.

8. **Graph your data.** Use the graph paper provided to make a graph of the class data, recording time for the steel ball to drop (in seconds) on the *y*-axis versus volume of sodium borate added on the *x*-axis. Draw a smooth curve through your data points.

Post-lab Questions

1. In Teflon tape, predict whether or not the polymer stands arranged randomly or in parallel. Explain your reasoning.

2. PFOA recently has been in the news in connection with Teflon and other fluorinated plastics. Use resources from the web and/or your instructor to answer these questions.

   a. What does PFOA stand for?

   b. List two current concerns about PFOA.

   c. How might the principles of Green Chemistry be used to address these and other concerns?

3. When a high-density polyethylene films is stretched, a "neck" can form. What happened to the arrangement of the polymer strands in this region of the polymer?

4. Materi-Bi is one example of the new biodegradable polymers that have come on the market in the last decade. Use the resources of the internet to answer these questions. Cite your sources.

   a. Identify two other commercially available biodegradable polymers.

   b. Besides their use in the collection of waste for community composting, what other applications do biodegradable polymers have? (Hint: have you ever had dissolving stitches to close a wound?)

5. Nylon is often drawn into fibers. What lengthwise and crosswise strength do you think would be important for a fiber? Explain your reasoning. Did your observation of nylon's properties match this expectation?

6. Polyurethane solutions A and B are sometimes injected directly into the walls of older homes and allowed to foam in place. Why?

7. Both polyurethane foam and Polyvinyl alcohol gel contain cross-links. What are cross-links? How do the strengths of cross-links in these two polymers compare to each other? Describe your evidence.

8. Describe the relationship between the viscosity of polyvinyl alcohol gel and the amount of sodium borate added. Explain the connection to cross-linking.

Name _____   Date _____

Lab Partner _____   Lab Section _____

Data Sheet—Experiment 23

Part I: Examination of Polymer Films

1. Properties of Teflon® tape:

2a. Initial properties of high-density polyethylene film:

2b. Properties of stretched polyethylene film:

3. Properties of Mater-Bi® film:

4. Observations of reaction of polymers with 3 M H_2SO_4.

Part II: Synthesis of Nylon

1. Properties of wet nylon strand:

2. Properties of dried nylon strand:

Part III: Synthesis of Polyurethane Foam

1. Observations as the reaction proceeds:

2. Properties of the final polyurethane foam:

Part IV: Synthesis of Polyvinyl Alcohol Gel

1. Properties of original polyvinyl alcohol solution:

2. Properties of the final gel:

Part V: Viscosity of Polyvinyl Alcohol Gels

Assigned volume of sodium borate _____

Drop time for steel ball (seconds) _____

Properties of the gel:

Plot the data from your class
on the graph to the right.

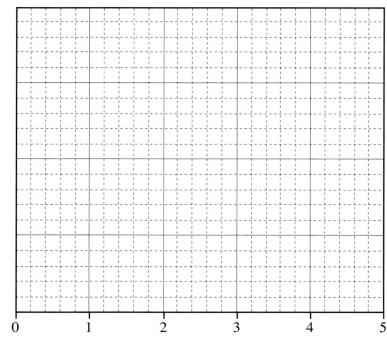

Why Do Plastics Get Sorted for Recycling?
Properties of Common Plastics

INTRODUCTION

Plastics have been an important part of our industrial society since Leo Baekeland invented the first synthetic polymer in 1907, a thermosetting resin called Bakelite that offered an alternative to cellulose resins. Today, plastics are found in everything from food and beverage containers, to furniture and electronic equipment. According to the most recent data available from the U.S. Environmental Protection Agency (EPA)[1], nearly 250 million tons of municipal solid waste was generated in the United States in 2008, of which just over 30 million tons, or about 12%, was plastics. Of all these plastics, a mere 2.1 million tons, about 7%, is recycled, well below rates for other materials such as aluminum (21%), glass (23%) and paper (55%). The majority of the waste is composed of six common plastics, and in this experiment you will examine the properties of these plastics and use a few simple tests to develop a classification and identification scheme.

Background Information

The plastics industry has adopted a code for packaging materials that can be used to identify the type of plastic in a plastic container. The idea behind the symbol is to make recycling easier by making the identification of the plastics easier. The symbol on the bottom of many containers is a triangle of arrows chasing each other with a number in the middle of the triangle.

The code for these symbols is as follows:

Number	Name	Abbrev.
1	Polyethylene terephthalate	(PET)
2	High-density polyethylene	(HDPE)
3	Polyvinyl chloride	(PVC)
4	Low-density polyethylene	(LDPE)
5	Polypropylene	(PP)
6	Polystyrene	(PS)
7	Other (includes composites and mixtures)	

---

[1] U.S. EPA, *Municipal Solid Waste Generation, Recycling and Disposal in the United States: Facts and Figures for 2008.* http://www.epa.gov/epawaste/nonhaz/municipal/msw99.htm

Compliance in labeling is voluntary, and not all plastics have an identification code symbol. Without code numbers these plastics are difficult to separate or classify by appearance. In this experiment, you will use four simple tests to aid in classification and identification of the six types of plastics. The first test will be to determine the relative densities of the plastics by checking to see whether the samples float or sink in three liquids of differing densities.

The second test will be to melt the plastic. All six of these common plastics melt reversibly, which means that when they are cooled, they harden and may regain their original properties. If a plastic sample does not melt, it is a thermosetting plastic. Thermosetting plastics, such as melamine (used in high-quality plastic dinnerware) or the plastic used in electrical components, do not melt cleanly and reversibly, but tend to char instead.

The third test, burning the plastic, must be performed only in a fume hood. All common plastics burn (some only if held directly in the flame), but with slightly different characteristics and different noxious fumes. The vapors given off from the burning plastic may have different properties depending on the plastic.

The last test, the copper-wire test, will be used to determine if a halogen, such as chlorine, is part of the polymer.

Overview of the Experiment

1. Obtain known samples of the 6 plastics.
2. Determine the relative densities of the samples.
3. Test each sample for melting.
4. Test each sample to see if it burns.
5. Test each sample with the copper-wire test.
6. Devise a classification scheme and a way to identify each plastic.
7. Test your scheme with some known samples provided by the instructor.
8. Test your scheme with some unknown samples that you brought from home.

Pre-lab Question

List five things you used today that are composed of plastic. If possible, locate the recycling symbol on each item and identify the type of plastic used to make it. Note that some items, such as electronics, do not commonly have the recycling symbol although they are made of plastics. Do you plan to recycle each item when it has served its purpose and you do not need it anymore? Which are easy to recycle, and which are more difficult?

EXPERIMENTAL PROCEDURE

I. Density Tests

Three liquids of differing densities will be used, as shown in the following chart.

Liquid	Density g/cm^3
1:1 95% Ethanol/ water	0.94
Water	1.0
10% NaCl in water	1.08

1. Obtain 3 test tubes. Pour about 5 ml of the 1:1 95% ethanol/water mixture in the first test tube and label it with its density (from the chart above). In the same fashion, put water in the second tube and 10% NaCl solution in the third tube. Be sure each tube is labeled.

2. Obtain two narrow strips of each of the 6 types of plastic. Cut *one* strip of each plastic into 3 small pieces.

3. For one of the plastics, place a piece into each of the three test tubes containing the density test liquids. Push each piece under the liquid surface with a glass stirring rod or the end of a pencil. If the sample floats, it has a density lower than that of the liquid. If it sinks, it has a density greater than that of the liquid. Record your observations on the data sheet.

4. In the same fashion, test each of the other plastics and record your observations.

5. Finally, use your observations to sort the six plastics by their densities, filling in the table and the ranking list on the data sheet.

II. Melt Test

1. Place a small sample of each plastic, one at a time, on the end of a metal spatula and hold the end of the spatula over a light blue microburner flame.

 STOP! Take great care when using an open flame. Long hair should be tied back, and extremely loose sleeves should be avoided. Pieces of hot molten plastic can cause burns if dropped onto your skin or clothing.

2. Heat slowly ("cook the plastic") and observe the plastic as it warms and finally melts. **Do not** heat strongly enough for the plastic to catch fire. Enter your observations in the data table.

3. Cool the sample and examine it for appearance and flexibility by bending it. Enter your observations in the data table.

4. The melted plastic can still be used for the tests that follow.

III. Ignition Test

 CAUTION! Toxic fumes are produced during the ignition test, so it is imperative that this part of the experiment be done in a fume hood.

1. Place a microburner or Bunsen burner in a fume hood. Light it and adjust to a small flame.

2. Place a large beaker of water in the hood.

3. Hold one end of a small strip of plastic in a pair of tongs, forceps, or pliers and place it directly in the flame. Observe the color of the flame and its characteristics. (Is a lot of smoke or visible vapor given off? Does the plastic continue to burn after it is removed from the flame?) Record your observations in the data table.

4. Test the vapors given off for acidic properties by holding a piece of <u>wet</u> litmus paper in the vapors from the burning plastic. If the paper turns red, acidic fumes are being formed as the plastic burns. Record your observations in the data table.

5. Extinguish the burning plastic by dropping it into the beaker of water.

6. Repeat steps 3–7 for each of the other plastics.

IV. Copper-Wire Test

 CAUTION! Toxic fumes are produced during the copper wire test, so this part of the experiment MUST be done in a fume hood.

1. Push the end of a 6-inch length of copper wire into a small cork.

2. Use the cork as a handle and heat the free end of the wire in a burner flame until the flame has no green color.

3. Touch the hot copper wire to the plastic you are testing and then return the wire end to the flame. A tiny bit of plastic should be picked up by the hot wire. Return the wire end to the flame. When the tip of the wire is put in the flame, watch for a slight flash of luminous flame. This indicates that you have correctly picked up a little bit of plastic on the wire.

4. Watch for the appearance of a green flame or green color in the flame when the plastic is heated in the flame. The green color indicates the presence of chlorine in the plastic.

V. A Puzzle for You to Solve

Devise a scheme for identifying the six plastics, based on the simple tests you have performed. The challenge is to find the minimum number of tests that will correctly identify the plastics if you are given any one of the plastics as an unknown. After you have devised your scheme, use it to identify at least two plastic samples that you have brought from home.

Finally, two "unknown" plastic samples will be given to you to identify. Use your scheme to determine their identities.

Post-lab Questions

1. Outline your scheme for identifying the six common plastics.

2. Suppose you had to add two other plastics to your scheme, polymethymethacrylate (density 1.18–1.20 g/cm^3) and poly-4-methyl-1-pentene (density 0.83 g/cm^3). Where would they fit in your scheme?

3. Polyethylene terephthalate (PET) is the most valuable waste plastic at the present time. Suggest a way to separate it commercially from other waste plastics.

4. Why are plastic recyclers very concerned about identifying the different polymers and not mixing them together?

5. Since waste plastic consists mostly of hydrocarbon compounds, it has been suggested that waste plastic could be used as fuel. Based on your observations in this experiment, do you think this is a reasonable suggestion? Would some be more dangerous to burn than others? Defend your answer.

Name _____   Date _____

Lab Partner _____   Lab Section _____

Data Sheet—Experiment 24

Unknown 1 number_____

Unknown 2 number_____

Part I: Density Tests (write in either "sinks," "floats," or "can't tell")

Type of plastic	1:1 Ethanol/H_2O density 0.94 g/cm^3	Water density 1.0 g/cm^3	10% NaCl solution density 1.08 g/cm^3
PET			
HDPE			
PVC			
LDPE			
PP			
PS			
Sample from home #1			
Sample from home #2			
Unknown 1			
Unknown 2			

Plastic Density Categories: (place each plastic into one of these categories)

Less than 0.94 g/cm^3	Less than 1.0 g/cm^3	Less than 1.08 g/cm^3	More than 1.08g/cm^3

Ranking of densities:

(lowest) _____ _____ _____ _____ _____ _____ (highest)

Parts II, III and IV: Melting and Ignition Tests

Plastic	Melting test	Ignition test	Copper-wire test
PET		Acidic smoke? yes/no _____	
HDPE		Acidic smoke? yes/no _____	
PVC		Acidic smoke? yes/no _____	
LDPE		Acidic smoke? yes/no _____	
PP		Acidic smoke? yes/no _____	
PS		Acidic smoke? yes/no _____	
Sample from home		Acidic smoke? yes/no _____	
Sample from home		Acidic smoke? yes/no _____	
Unknown 1		Acidic smoke? yes/no _____	
Unknown 2		Acidic smoke? yes/no _____	

Identification of Unknown Samples

Sample from home #1: Description _____ Identity _____

Sample from home #2: Description _____ Identity _____

Unknown 1 number _____   Identity _____

Unknown 2 number _____   Identity _____

What Drugs are in an Analgesic Tablet? Identification by Thin-Layer Chromatography

INTRODUCTION

In this experiment, you will utilize thin-layer chromatography to identify the analgesic compound(s) present in an over-the-counter (OTC) painkiller preparation. OTC analgesic preparations are widely sold and used in the United States. Although extremely useful and effective, they are also powerful drugs that can be very harmful if misused. Because such analgesics are so commonly available in homes, there is a danger of accidental overdose by young children who discover the innocuous-looking white pills. In cases of accidental poisoning, there may be need for quick identification of what is in an unknown tablet. This experiment demonstrates one such method.

Analgesic compounds are discussed in Sections 10.3 and 10.4 of *Chemistry in Context*. Aspirin (acetylsalicylic acid) is by far the most common. Other common non-prescription analgesics include acetaminophen (Tylenol) and ibuprofen. Caffeine is sometimes added to analgesic tablets to overcome drowsiness. Tablets also contain binders such as starch and sugars. "Buffered" aspirin contains a base such as magnesium hydroxide or calcium carbonate to neutralize the acid.

Background Information

Chromatography is a technique used to separate and identify individual components in a mixture. You may have experience with this technique if you have performed Experiment 2. Its name, from the Greek for "writing with color", suggests the earliest applications of chromatography that dealt with mixtures of colored substances such as plant pigments. In its simplest form, a chromatographic set-up consists of an immobilized substance (called the "stationary phase") over which a gas or liquid (the "mobile phase") moves. Components of a mixture injected into the mobile phase can be separated based on their tendency to move along with the mobile phase rather than be attracted to the stationary phase. A variety of chromatographic techniques have been developed, and all chromatographic techniques make use of the fact that components of a mixture (either gaseous or in solution) tend to move at different speeds based on the differing attractions of the components to the stationary phase compared to the tendency of the components to remain in the mobile phase (either gas or liquid).

Thin-layer chromatography (TLC) is one of the easiest chromatographic techniques. A thin layer of a suitable solid substance is coated on a sheet of glass, plastic or aluminum. By immersing one edge of the sheet in an appropriate liquid solvent, the solvent is drawn up the sheet by capillary action, and the compounds of interest are carried along at differing rates. This is commonly called "developing" the plate.

Unless the compounds of interest are colored, some method is necessary to make them visible. One method is to expose the developed plate to some compound that will react chemically with the spots to make them visible. Another method is to examine the plate under ultraviolet light to see if any of the components are fluorescent. An ingenious alternative to that is to incorporate a fluorescent dye into the plate so that when the plate is examined under a UV lamp, the plate will glow (fluoresce) everywhere *except* where the component spots are.

Overview of the Experiment

1. Prepare a developing container.
2. Spot known and unknown samples on a chromatographic sheet.
3. Develop the chromatogram.
4. Calculate R_f values.
5. Identify components in unknown sample.

Pre-lab Question

Why do you think it is necessary to maintain a saturated vapor atmosphere in the container used for TLC development? What do you predict might happen if this were not done?

EXPERIMENTAL PROCEDURE

1. Obtain a large beaker (400–600 mL) and a piece of plastic wrap or aluminum foil to cover it. Alternatively, a wide-mouthed jar with screw-cap lid may be used. The beaker or jar should be large enough so that the chromatographic sheet can lean against one side (see *Figure 25.1*).

2. Add enough of the solvent mixture (containing 25 parts ethyl acetate, 1 part ethanol, and 1 part acetic acid) to give a thin layer of solvent in the bottom of the container. To provide an atmosphere saturated with solvent inside the container, place a piece of filter paper around the inside surface of the container, extending into the solvent (see *Figure 25.1*). Then cover the container with the plastic wrap, foil, or screw cap and set it aside while preparing the chromatographic sheet (steps 3–6).

3. Obtain an unknown analgesic tablet (or a portion of one). Using a mortar and pestle, crush a portion of a tablet to a fine powder. (Alternatively, this can be done in a small test tube with a glass rod, but it must be done carefully to avoid breaking the test tube.) Add a few milliliters of methanol to the powder and stir; then allow the mixture to settle for a few minutes.

4. Obtain a piece of plastic-backed chromatographic sheet. Unless you are told otherwise, this is a special kind of TLC sheet that contains a fluorescent dye to help make the results more visible. The white coating may flake off if the sheet is not handled carefully. Handle it only

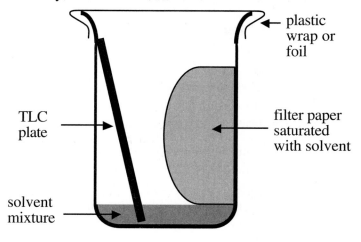

Figure 25.1 The assembled TLC developing chamber.

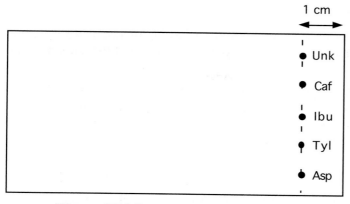

1 cm

• Unk

• Caf

• Ibu

• Tyl

• Asp

Figure 25.2 Layout of TLC sheet

on the edges. The backing is flexible, but the sheet should not be bent excessively. Using a pencil (not a pen), draw a *very light* line across the sheet in the short dimension about 1 cm from one end (*Figure 25.2*). Then make 5 small light marks at 1-cm intervals along the line. These are the points at which the samples will be spotted. (Your instructor may suggest different spacing, depending on the size of TLC sheet and the number of spots used.) Label each point with a penciled number or abbreviation and record this code on the data sheet.

5. Samples will be spotted onto the chromatographic sheet, using glass capillary tubes drawn down to a very small opening. (The tubes are quite fragile.) Before spotting the sheet, you should practice putting *very small* spots of solution on a piece of scrap paper. To do this, dip the tip of a capillary into a solution, then very gently touch the tip to the paper for a brief moment. The spots should be as small as possible in order to minimize tailing and overlapping when the chromatographic sheet is "developed." (If a more intense spot is desired, it is preferable to let the spot dry and re-spot in the same location.)

6. Solutions of the 3 analgesic compounds plus caffeine will be available in the lab with a glass capillary tube in each. Carefully place small spots of the four solutions at four pencil marks. (**Optional:** your instructor may ask you to add another spot with a mixture of all four compounds.) Finally, spot a sample of the clear solution from your unknown tablet onto the chromatographic plate, using a separate glass capillary tube. Allow the solvent to evaporate. A hair dryer or oven can be used to speed up the drying process.

7. When the spots are dry, place the sheet in the developing container (*Figure 25.1*). Check to be sure that the bottom edge (near the spots) is in the solvent but that the spots are above the solvent. Also be sure that the filter paper does not touch the chromatographic sheet. Then cover with plastic wrap, foil, or a screw-cap lid and watch carefully. The liquid will slowly move up the TLC sheet by capillary action.

8. When the front edge of the liquid has moved within about 1 cm of the top of the plate, remove the plate from the TLC developing chamber. *Immediately*, while the sheet is still wet, draw a pencil line on the sheet to show the top edge of the liquid. Then lay the sheet on a clean surface in a fume hood or well-ventilated area and allow the solvent to evaporate until the sheet appears dry.

9. The spots are unlikely to be visible to the naked eye, but they should be quite visible when viewed under an ultraviolet (UV) lamp. It may be necessary to do this in a location with subdued light. While observing under the UV lamp, draw a light pencil line around each spot.

 CAUTION! UV radiation is harmful to your eyes. Do not look directly at the UV lamp.

10. A more elegant way to interpret the results is to calculate a *retention factor* (R_f) for each spot. The retention factor is the ratio of the distances traveled by the component and the solvent.

$$R_f = \frac{\text{distance traveled by the substance}}{\text{distance traveled by the solvent}}$$

a. Obtain a short ruler marked in centimeters and millimeters. For each spot, carefully measure the distance (*in millimeters*) from the starting line to the solvent front and also from the starting line to the middle of the spot. Some judgment may be necessary in choosing the middle of the spot. Record these numbers on the data sheet. If more than one spot is present for the unknown, make these measurements for each spot.

b. Calculate the R_f value for each spot, dividing the distance to the center of the spot by the distance to the solvent front. This number should be less than 1, since the spot did not move as far as the solvent. Record this ratio on the data sheet, writing it as a decimal fraction rounded to two digits after the decimal point.

Post-lab Questions

1. Suggest possible advantages and disadvantages of using a longer (taller) TLC sheet.

2. Why do you think it was important to use a very small amount of sample when spotting the plate?

3. The relative movement of components is controlled partially by the polarity of the molecules. The TLC sheet is coated with a highly polar substance, whereas the solvent mixture has a much lower polarity. From your chromatographic results, predict the relative polarities of aspirin, acetaminophen, ibuprofen, and caffeine. Explain your reasoning.

4. Do you expect that changing the solvent will change the R_f value for a given component? Explain your reasoning. *Yes be cause every solvent is different and would react with the samples different tlc*

5. If two components have an identical R_f value, does this mean they necessarily have the same structure? Explain why or why not. *No They wouldn't necessarily have the same structure - Because something is similar Does not make it the same*

6. In an effort to identify an unknown, some students obtain the TLC plate shown at the right.

a. How many compounds were in Sample A? In Sample B? In the unknown? *A -2   B-1*

b. A student concludes that the unknown is the same as Sample B because of the number of spots. Is this a valid conclusion? Explain. *No Because the compounds were drawn up at different rates*

c. A second student concludes that the unknown is the same as Sample A because they both have spots with R_f values of 0.3. Is this a valid conclusion? Explain. *No . Because sample A has 2 compounds in it and the unknown has 1 compound it may be the same in both*

d. Propose the most reasonable conclusion you can for the identity of the unknown. Explain your reasoning. *The unknown may have the same components of A because 1 component of A was at the same level as the unknown*

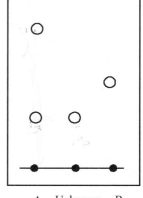

A   Unknown   B

How is Aspirin Made?

INTRODUCTION

This experiment is a representative example of a chemical synthesis; in this case, it is the synthesis of a common analgesic drug. Of all the drugs discussed in Chapter 10 of *Chemistry in Context*, aspirin can be made most easily. It is prepared from salicylic acid by a reaction known as esterification. You may explore two methods for synthesizing aspirin: (1) a conventional method very similar to the original synthesis of aspirin carried out by Felix Hoffman in 1893 in which a hotplate is used to heat the reaction, and (2) a similar reaction where microwave irradiation is used.

Background Information

Hoffman's method for producing aspirin involved reacting salicylic acid with acetic acid in the presence of sulfuric acid as a catalyst.

For the conventional synthesis, instead of using acetic acid, you will use acetic anhydride, a compound that produces acetic acid when it reacts with water. Acetic anhydride is prepared by combining two acetic acid molecules.

Acetic anhydride works better in the traditional aspirin synthesis than acetic acid and requires less catalyst. The reaction is as follows:

For the microwave synthesis, you will also use acetic anhydride, but no sulfuric acid catalyst is required. Microwaves are a form of electromagnetic radiation with frequencies in the range of 300 to 300,000 MHz. Household microwaves are used to cook food, but microwave sources can also be used in the laboratory to heat reactions. Microwaves interact with polar molecules and ions, causing them to rotate rapidly and heat up. Typically, reactions conducted in the microwave take place more quickly and with reduced consumption of energy compared to reactions heated in a conventional way. Thus, microwave chemistry is often considered to be a "green" technique.

After the aspirin is synthesized, it must be isolated and purified. The impurities in the crude aspirin are unreacted starting materials (salicylic acid, acetic acid or acetic anhydride), catalyst (sulfuric acid), and byproducts. Fortunately, most of these compounds are very soluble in water and easily washed away from the mixture during the isolation of the crude (unpurified) aspirin. However, salicylic acid and aspirin are only sparingly soluble in water, and any unreacted salicylic acid may be a contaminant in the crude product mixture. A technique known as **recrystallization** will be used to separate the aspirin from the unreacted salicylic acid. This technique involves dissolving the crude mixture in a suitable warm solvent. The solution is cooled to a point at which one of the components of the mixture crystallizes out of solution while the other component (the more soluble one) remains dissolved. The crystallized component can then be collected by filtration.

You can examine the purity of your products by measuring the melting points. Every pure compound has a characteristic temperature at which it melts, and pure compounds usually melt sharply within a range of 1° C. The melting point gives an indication of the purity of a compound because any impurities will lower the melting temperature and increase the melting range. Therefore, a melting point lower than the accepted value indicates either that the compound is impure or that it is not what you think it is.

Note: This experiment requires more equipment and more steps than most of the previous laboratory exercises, so a list of required materials has been provided for you at the beginning of each section. It is important for you to read the instructions carefully and be sure you understand them before proceeding.

Overview of the Experiment
1. Collect the necessary equipment and chemicals.
2. Combine salicylic acid, acetic anhydride, and sulfuric acid, and heat in a water bath.
3. Combine salicylic acid and acetic anhydride, and heat in a microwave.
4. Collect the crude products by vacuum filtration.
5. Purify the products by recrystallization.
6. Compare the properties of aspirin (crude and purified) with salicylic acid.

Pre-lab Questions
1. Give the range of wavelengths, in meters, for microwave radiation. How does this compare to visible light?
2. Sulfuric acid is used as a catalyst for the conventional reaction. Define the term "catalyst". This catalyst is not added to the microwave reaction, but the reaction is still acid-catalyzed. What is the source of the acid?

EXPERIMENTAL PROCEDURE

STOP! Eye protection is absolutely essential for this experiment to keep the corrosive acids out of your eyes. If you spill acid on yourself, wash it off with large amounts of water. Keep the portion of you that came into contact with acid under running water for at least 5 minutes, and notify your laboratory instructor. If you spill acid on your workspace, notify your lab instructor who will be able to properly clean up the spill. If you are allergic to aspirin, notify your lab instructor before proceeding.

I. Synthesis of Aspirin
<u>Materials required:</u>
- 2 g salicylic acid
- 3 mL acetic anhydride
- 10 drops concentrated sulfuric acid
- 2 mL distilled water
- 250-mL beaker for a hot water bath
- 125-mL Erlenmeyer flask
- 10-mL graduated cylinder
- glass stirring rod
- crushed ice and beaker for ice bath
- small Büchner or Hirsch funnel with filter paper
- filter flask, tubing and filtering gasket

1. Half fill a 250-ml beaker with hot water and put it on a hot plate. This will serve as a hot-water bath.

2. Weigh 2.1 grams of salicylic acid (0.015 mole) into a <u>dry</u> 125-mL Erlenmeyer flask. (It is crucial that the flask be dry. Any traces of water will interfere with the synthesis.) Record the appearance of the salicylic acid.

CAUTION! Sulfuric acid and acetic anhydride are corrosive, and acetic anhydride has irritating fumes. Perform steps 3-8 in a fume hood.

3. In the fume hood, carefully measure 3 ml of acetic anhydride in a <u>dry</u> 10-ml graduated cylinder.

4. Add the acetic anhydride to the salicylic acid. Swirl to mix the liquid with the solid.

5. Carefully add 3 drops of sulfuric acid to the mixture.

6. Put the flask into the hot-water bath (which should be approximately boiling). Heat for 10 minutes. If necessary, stir the mixture with a glass rod to break up any lumps. Be sure you have heated for at least 5 minutes past the point where most of the solid has dissolved.

7. Remove your flask from the hot water. Add 2 mL of distilled water, a few drops at a time, with continuous swirling. *Do not add the water all at once.* The water is reacting with leftover excess acetic anhydride. Solid crystals of aspirin may start to form at this point.

8. Then add 10 grams of crushed ice to your flask and swirl. You should have a white solid product (aspirin) at this point. If not, consult your instructor before proceeding, especially if you see an oily liquid phase rather than a solid product. (Scratching the inside of the flask below the liquid level with a glass stirring rod will usually cause the oil to crystallize. This may take several minutes. Alternatively, you can reheat the flask in a hot-water bath until all the oil disappears, then transfer the flask to an ice bath and swirl.)

9. Return to your regular workbench. Cool your flask in an ice/water bath that you construct in a beaker. Use roughly a 2:1 mix of ice and water. (An optimal ice/water bath is mostly ice but with enough water to produce a slurry of ice and water.) Continue to cool the flask for 5–10 minutes with occasional swirling or stirring.

10. While the reaction mixture is cooling, prepare a vacuum filtration setup such as the one shown in *Figure 26.1*. Your equipment may look slightly different from this. Your instructor will demonstrate how to use it. (It may be possible for several teams to share one setup.)

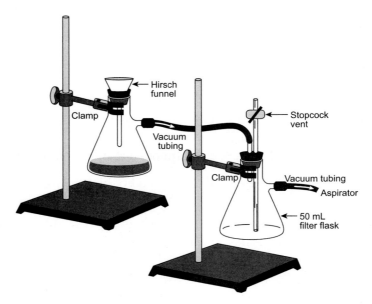

Figure 26.1 Diagram of the filtration setup

11. Attach one end of the tubing to the side arm of the filter flask and the other end to the aspirator side arm on a water faucet.

12. Attach a small Büchner funnel or Hirsch funnel to the filter flask and put a piece of filter paper in the funnel. The filter paper should lay flat and just cover all the holes in the bottom of the funnel.

13. Wet the paper with a few drops of water.

14. Turn the water full on. (Depending on the sink setup, you may need to add some sort of splash arrester in order to avoid being drenched.) When the water is on, there should be enough suction so that the filter paper is held firmly to the funnel.

15. Quickly pour the cooled slurry of product onto the center of the filter paper. If your setup is working correctly, the water should be sucked through the funnel, leaving a white pasty product on the filter paper.

16. Rinse the reaction flask with 2–3 mL of *ice-cold* water and pour the rinses through the funnel.

17. Keep the suction on for several minutes so that air is drawn through the solid until the product is nearly dry. Spreading the aspirin evenly on the filter paper will help speed the drying.

18. When the product is dry, detach the tubing from the side arm and then shut off the water. (Do NOT shut off the water first.)

19. Transfer the solid to a preweighed watch glass by inverting the funnel over the watch glass and tapping gently. Reweigh and record the mass. You should have about 2 grams of product.

II. Microwave Synthesis of Aspirin[1]
<u>Materials required</u>
- 0.7 g salicylic acid
- 1.4 mL acetic anhydride
- 25 mL distilled water
- 10-mL graduated cylinder
- 50-mL beaker
- glass stirring rod
- 50-mL Erlenmeyer flask
- microwave oven inside a fume hood
- filtration apparatus from Part I, with a clean funnel and new filter paper
- ice bath from Part I

Note that these instructions are for a 1200-W microwave oven. If the microwave available for your use differs in power, your instructor may give you a modified heating program. You should follow the instructions given in the lab if they differ from these.

1. Weigh 0.7 g salicylic acid using a laboratory balance and carefully pour it into a 50-mL beaker.

2. Measure 1.4 mL acetic anhydride using a graduated cylinder and add that to the salicylic acid in the beaker.

3. Stir the mixture and place it in the microwave oven. Heat the reaction mixture at 80% power for 2 minutes. Then, remove the reaction mixture and stir it gently.

 CAUTION! The microwave should be placed in a fume hood, and the sash on the fume hood should be pulled down while the microwave is operating. Use caution (and perhaps a pair of gloves) when removing your beaker from the microwave as it may be very hot to the touch.

4. Heat the reaction for another 2 minutes at 80% power. Stir it gently and then allow it to sit undisturbed on the counter in the fume hood. Crystals should begin forming after about 10

---

[1] adapted from: Montes, I.; Sanabria, D.; Garcia, M.; Castro, J.; Fajardo, J. A Greener Approach to Aspirin Synthesis Using Microwave Irradiation. *J. Chem. Ed.* **2006**, *83*, 628-631.

minutes. If no crystals form within 20 minutes, gently scratch the inside of the glass beaker with a glass rod. If crystals still do not form, consult your instructor.

5. Once your reaction is cooled to room temperature, place the beaker in the ice bath to cool it further. Be sure to not let your beaker tip over in the ice bath or you will lose your product. While your reaction is cooling, place 25 mL of distilled water into a 50-mL Erlenmeyer flask, and place the flask into the ice bath to cool the water.

6. Once the reaction has completely cooled, add the cold water to the crystals in the beaker and stir. This will dissolve any unreacted acetic anhydride, but will not dissolve the crystallized aspirin.

7. Filter the aspirin crystals as described in Steps 10-19 in Part I. You should end up with about 0.8 g of product.

III. Purifying the Aspirin by Recrystallization
<u>Materials required:</u>
- crude reaction mixture from Part I
- hot water bath and ice bath from Part I
- 50-mL Erlenmeyer flask
- 6 mL ethanol
- 10 mL distilled water
- filtration apparatus from previous parts of experiment

At this point, the aspirin you obtained from each procedure is somewhat wet and probably still contaminated with some salicylic acid. You will further purify the crude mixture obtained from the conventional synthesis by a technique known as recrystallization. The recovered crystals from the microwave synthesis should be fairly pure and will be analyzed directly in Part IV.

1. Save a small amount of each crude aspirin product for later comparison. Place the rest of it into a 50-mL Erlenmeyer flask. Add 6 mL ethanol**.** Heat this flask in a hot-water bath on a hot plate, preferably in a fume hood. The solid should all dissolve.

 CAUTION! Ethanol is flammable. Be sure that there are no open flames anywhere in the lab before starting this part of the experiment.

2. Add 10 mL distilled water to the flask and reheat in the hot-water bath until all of the solid again dissolves.

3. As soon as the solid has dissolved, take your flask back to your regular workstation and let it cool slowly to near room temperature. Do <u>not</u> cool rapidly with ice water. Crystals of product should form during the cooling. If not, consult your instructor.

4. Finally, cool the flask for 5 more minutes in an ice/water bath with occasional stirring. (**Warning:** Do not let the flask tip over in the ice/water bath. If it does, you will lose your product!)

5. Collect the product by suction filtration as described above.

6. Spread your product out to dry on a watch glass or paper towel.

At this point, your instructor will tell you if you can leave the product to dry thoroughly until the next laboratory period, and complete the experiment then. If not, you can let it dry for a short time while you clean up your equipment and answer the post-lab questions at the end.

7. Carefully observe and describe this purified aspirin product. (This can be done while it is drying or afterward.) Does it appear different from the salicylic acid that you started with? If possible, examine a few crystals under a microscope or with a magnifying lens and observe the crystal shapes. Also examine the crude product. Record your observations in the data table.

8. Finally, determine the mass of purified aspirin by weighing it in a preweighed or "tared" vial or beaker. Many laboratory balances permit taring, which means that the balance can be set to read zero when the empty container is in place; thus, the measured mass with aspirin added is simply the mass of aspirin.

IV. Analysis of Product Purity

One of the classic ways to identify and determine the purity of an organic chemical compound such as aspirin is to measure its melting point. Your laboratory may have a special apparatus for determining the melting point of organic compounds, or an alternate procedure may be used. In any case, your instructor will demonstrate how to obtain this measurement.

1. Measure the melting temperatures of salicylic acid, pure aspirin (if available), your crude and recrystallized aspirin synthesized by conventional heating, and the product of the microwave reaction. Record the melting temperatures as ranges, recording the temperature at which the compound first begins to melt, and the temperature at which the compound is completely melted.

Another way to assess the purity of your aspirin would be to analyze it using thin layer chromatography, as described in Experiment 25. Your instructor may provide you with the necessary materials to analyze your synthesized aspirin against standard solutions of aspirin and salicylic acid. Alternatively, if you perform this experiment before Experiment 25, you may be able to save your product to analyze alongside the analgesic tablets.

Post-lab Questions

1. Draw the structures of salicylic acid and aspirin. Circle and identify the functional groups present in each compound.

2. The equation for the synthesis reaction is given in the Introduction. Copy this equation and indicate on the structures which bonds are broken in the reaction and how the resulting broken pieces pair up to give the products.

3. What impurities do you think are present in your crude aspirin product? Where do those impurities go in the recrystallization process?

4. What safety, environmental, and economic considerations would be important to a company manufacturing 1 million pounds of aspirin each year by this procedure? Which ones do you think are most important? Why?

5. The aspirin synthesized in this experiment should *never* be taken home for medicinal use. Why not?

6. It should be noted that the conventional reaction will not take place if a catalyst is not added to the reaction mixture. Yet, good yields can be obtained in the microwave without a catalyst. Speculate why.

7. Look at the Principles of Green Chemistry in the inside cover of your book. Which principles does the conventional synthesis follow well? Which does the microwave synthesis follow? Which synthesis do you think is "greener"? Provide support for your argument.

Name _____   Date _____

Lab Partner _____   Lab Section _____

Data Sheet—Experiment 26

Physical observations of salicylic acid:

Part I – Conventional Aspirin Synthesis

Mass of salicylic acid used (grams): _____

Mass of crude aspirin (grams): _____

Mass of purified aspirin (grams): _____

Physical observations of the crude aspirin product:

Physical observations of the purified aspirin product:

Does your purified aspirin appear to be purer than crude aspirin? What is your evidence?

You started this synthesis with 0.015 mole of salicylic acid. Do the following calculations and show your work:

a. If the reaction worked perfectly, how many moles of aspirin would you make? (Look at the balanced equation for the synthesis reaction.)

b. Convert this answer into grams of aspirin. To do this, multiply the number of moles of aspirin by the molar mass of aspirin (180 g/mole).

c. Compare your actual mass of purified aspirin to the predicted mass above. What is your percent yield?

$$\text{percent yield} = \frac{\text{actual mass}}{\text{predicted mass}} \times 100\%$$

d. Suggest two possible reasons why your percent yield is less than 100%.

Part II – Microwave Aspirin Synthesis

Mass of salicylic acid used (grams): _____

Mass of recovered aspirin (grams): _____

Physical observations of the crude aspirin product:

You began the microwave synthesis with 0.005 moles of salicylic acid.   Repeat the yield calculations described above for this synthesis. Show your work.

Part IV – Melting points

Melting point of salicylic acid: Melting starts:_____Melting complete:_____

Melting point of pure aspirin: Melting starts:_____Melting complete:_____

Conventional synthesis

Melting point of purified aspirin: Melting starts:_____Melting complete:_____

Melting point of crude aspirin: Melting starts:_____Melting complete:_____

Microwave synthesis

Melting point of aspirin: Melting starts:_____Melting complete:_____

Impurities tend to lower the melting-point temperature of a solid and make it melt over a range of temperatures rather than at one sharp temperature. The published melting-point temperatures are 159°C for salicylic acid and 135°C for aspirin. Discuss your melting point data compared to these standard values. How pure are your products?

How Much Fat Is in Potato Chips and Hot Dogs?

INTRODUCTION

Fat, sugar, and salt are three food components that worry nutritionists because Americans and others in developed countries tend to consume too much of these food ingredients. Excess consumption of these food components can lead to problems such as obesity and heart disease, and many doctors and journalists believe that we are in a public health crisis due to our eating habits. In this experiment, you perform two extractions to measure the amounts of fat in foods that are a common part of our modern American diet.

Two approaches will be used to determine the fat content of foods. In fat-coated foods such as potato chips, the fat can simply be dissolved out of the food using a solvent. However, another approach must be used for meat products (such as hot dogs) because the fat is bound up in animal tissue (protein). Therefore, the meat must be chemically treated to break down the protein and other tissues in order to liberate the fat. Since the liberated fat can float on water, it can be easily removed and weighed.

I. Fat in Potato Chips

This study is conveniently done as a class project in which each student group analyzes a different brand of potato chips or other snack chips. Alternatively, each pair of students can do one measurement on each of two kinds of chips, then pool their results with those of other students. Your instructor will indicate how the class assignment will be structured. The essential procedure is very simple: Petroleum ether (a commercial mixture of hydrocarbons that is widely used as a solvent) is mixed with ground-up chips to extract the fat. After separating the solvent mixture from the chips, the petroleum ether is allowed to evaporate, leaving behind the fat that can then be weighed.

II. Fat in a Hot Dog (or other meat sample)

The procedure given here is written for hot dogs, but it can be used for any kind of meat sample. Your instructor may divide up your class and have different student groups analyze various hot dog samples. For example, your class may wish to investigate whether different brands of hot dogs differ significantly in fat content, whether chicken or turkey hot dogs differ from beef and pork hot dogs, whether hot dogs and hamburgers differ significantly in fat, whether "low fat" products are really as claimed, and how vegetarian hot dogs compare to meat hot dogs in their fat content.

The experimental procedure utilizes a special solution identified simply as "protein liquefying reagent." This is a highly alkaline mixture containing sodium hydroxide, sodium salicylate, potassium sulfite, isopropyl alcohol, and water.

Overview of the Experiment

1. Obtain two weighed samples of potato chips or other snack chips.
2. Grind up the samples and mix them with petroleum ether to extract the fat.
3. Separate the petroleum ether from the potato chips by filtration into a weighed beaker.
4. Evaporate the petroleum ether on a steam bath or hot plate.
5. Weigh the beaker and calculate the mass of fat by difference.
6. Obtain a sample of ground hot dog or other meat.
7. Determine the masses of two separate hot-dog samples in test tubes.
8. React the hot-dog meat with protein liquefying reagent.
9. Cool and centrifuge the samples.
10. Remove the fat layer from each hot-dog sample and weigh it.
11. Calculate the percent fat in your chips and hot dog.

Pre-lab Question

Predict the relative fat content in all of the samples that your class will analyze.

EXPERIMENTAL PROCEDURE

Part I – Fat in Potato Chips

1. Obtain two small beakers that are clean and dry. Label them with your initials and identifying numbers (1 and 2).

2. Take a bag of chips, the two beakers, and this lab manual or the data sheet to a laboratory balance. Accurately weigh out two samples of chips of about 2–3 grams each. This is done most conveniently by using small sheets of paper or small plastic weighing dishes. Record the mass of the first paper or dish to the nearest 0.01 gram, add 2–3 grams of chips, and again record the mass to the nearest 0.01 g. Alternatively, your instructor may show you how to tare the dish for direct weighing of the chips. Weigh the second sample in the same manner. Keep track of which sample is which.

   Note about weighing: It is important to check the balance **each** time it is used to be sure it reads 0.00 g when there is nothing on the balance pan.

3. Also weigh the two beakers and record their masses to the nearest 0.01 gram in the appropriate places in the data table. (First check to be sure that the balance reads 0.00 g when empty.)

4. Put chip sample #1 in a clean porcelain mortar and use a pestle to crush the chips into very small pieces. (Because of the fat, the mixture may be rather sticky.)

5. Add 15 mL of petroleum ether to the mortar and grind the mixture thoroughly.

   CAUTION! Petroleum ether is a gasoline-like solvent that is *extremely* flammable. It is absolutely essential that no open flames be present anywhere in the laboratory during this experiment. Under no circumstances should the heating be done with a Bunsen burner.

6. Prepare a glass funnel with folded filter paper (as shown by the instructor). Mount it over weighed beaker #1. Using a spatula, carefully transfer the potato-chip mixture into the filter. Try to get most of the mixture into the filter.

7. In order to rinse out any remaining fat in the mortar and on the chip mixture, add 5 mL more petroleum ether to the mortar, stir it around with the pestle, then pour it onto the chip mixture in the filter.

8. Repeat with another 5-mL rinse of the mortar and the chip mixture in the filter.

9. Repeat steps 4–8 with the second sample of chips using the same mortar and pestle but a fresh piece of filter paper and weighed beaker #2.

10. When both mixtures have finished filtering, the next step is to remove all of the petroleum ether by evaporation. There are two ways this can be done.

    a. You can leave the beakers in a fume hood until the next day or the next lab period, by which time the solvent will have evaporated.

    b. A much faster way is to place the beakers on a steam bath or hot plate <u>in a fume hood</u>. Leave them for about 15 minutes or until all of the petroleum ether has evaporated. Then remove the beakers from the steam bath or hot plate, carefully wipe the outsides to remove all water, and let them cool for a few minutes.

11. Reweigh the beakers and record the masses to the nearest 0.01 gram. (First check to be sure the balance reads zero when empty.)

Part II – Fat in a Hot Dog

1. Put approximately 75 mL of water in a 250-mL beaker. Place the beaker of water on a hot plate and heat the water to between 80° and 90°C.

2. While the water is heating, obtain two centrifuge tubes and label them #1 and #2. Put the labels high on the tubes near the top. Take the tubes along with the lab book or the data sheet to a laboratory balance. Check to be sure that the empty balance reads exactly 0.00 g, then weigh the test tubes and record their masses to the nearest 0.01 gram.

3. Put between 2 and 2.5 grams of ground hotdog meat in each test tube and reweigh the tubes. Record the mass of each tube + hot dog sample to the nearest 0.01 gram.

4. Add about 5 mL of the protein liquefying reagent to each hotdog meat sample.

 CAUTION! Do not allow the protein liquefying reagent to contact your skin. It is highly caustic and can cause serious skin damage. If you do get it on yourself, immediately wash your skin with copious amounts of water and notify your instructor.

5. Put the test tubes in the beaker of water and heat the beaker **on a hot plate** (NOT a Bunsen burner) until the reagent in the tube starts to boil (about 80º C). Maintain the temperature of the water so that the contents of the test tubes boil for 10 minutes. Do not leave the beaker and test tubes unattended. The reaction must be monitored to ensure that liquid does not boil out of the test tube. In addition, the vapor from the protein liquefying reagent is extremely flammable so care must be taken.

6. After the mixture has boiled for 10 minutes, it should be dark brown with some yellow fat floating at the top. Remove the test tubes from the hot water, stand them in a 250-mL beaker, and let them stand until they are cool enough to handle. Do not let them cool completely because the fat may turn from liquid to solid.

7. Label two small containers (vials, small beakers, test tubes, or watch glasses) as "sample 1" and "sample 2" and then weigh them to the nearest 0.01 gram. (Remember to check first to be sure the balance reads 0.00 g when empty.) Record the masses on the data sheet.

8. Put the two test tubes in a centrifuge, placing them directly opposite each other. Other students may add pairs of test tubes at the same time. Centrifuge the mixtures to completely separate the fat from the rest of the solution. The fat should be floating on top.

9. Use a Pasteur pipet (glass tube with a long, thin stem and a small rubber bulb attached to the top) to slowly transfer the top fat layer from test tube #1 to the correct preweighed container. Be very careful to remove all of the fat but none of the brown liquid. Work slowly; it takes patience to do this correctly so that only the fat is removed. If you accidentally suck up some of the brown liquid, it can usually be removed from the bottom of the container of fat by using the Pasteur pipet.

10. Repeat the procedure for the other sample.

11. Weigh the containers of fat to the nearest 0.01 gram and record their masses in the data table.

Clean-up

Discard the filter paper and chip mixtures from Part I in a designated container. The instructor will specify what to do with the mortar and pestle (which should be nearly clean) and the beakers (which are quite greasy). The "brown protein liquid" remaining in the test tubes from Part II should be deposited in an appropriate waste container. Follow instructions for disposal of the fat in the vials.

Calculations

Part I

For each sample, calculate by subtraction the mass of chips and the mass of fat. Then calculate the percent fat in the chips:

$$\% \text{ fat} \ = \ \frac{\text{mass of fat}}{\text{mass of chips}} \ \times \ 100$$

Part II

By subtraction of masses in the data table, calculate the mass of hot dog and the mass of fat for each sample. The percent fat is then calculated from the relationship:

$$\% \text{ fat} \ = \ \frac{\text{mass of fat}}{\text{mass of hot dog}} \ \times \ 100$$

Sharing Your Results

Finally, share your results with others in your class or lab section. For Part I it will probably be advantageous to report your two results separately rather than as an average. If others analyzed the same brand, you can see how closely you agree. Examining the class data as a whole, you can look for differences and possible generalizations about different types of snack chips.

When you are finished with the calculations for Part II, share your results with the rest of the class or lab section. If different brands of hot dogs were analyzed, how do the fat contents compare? Can you draw any conclusions about relative fat content of different kinds of foods?

Post-lab Questions

Part I: Fat in Potato Chips

1. Looking at the class results, can you make any generalizations? For example, do potato chips have more or less fat than tortilla chips?

2. If your class analyzed any "low-fat" or "no-fat" products, do the results support this claim? If not, suggest a reasonable explanation.

3. Potatoes are naturally low in fat, so where does the fat in potato chips come from? How are low-fat chips different?

4. A snack pack of potato chips or other kinds of chips holds 1 ounce (28 grams) of chips. Based on your data, how much fat is present in a snack pack? In such a bag of chips, there are about 15 grams of carbohydrate and about 1 gram of protein. Given that carbohydrates and protein provide about 4 Calories of energy per gram and fats provide about 9 Calories per gram, what is the total energy equivalent (in Calories) of a 1-ounce bag of chips? What percent of the Calories are from the fat? *Show your calculations.*

Part II: Fat in Hot Dogs

1. If your class analyzed different kinds of hot dogs or other meats, what can you conclude about the fat content in these products? Are some products significantly lower in fat than others?

2. Why did hot dogs require a different fat extraction procedure than the chips did? Which of the two procedures do you think would work best when analyzing the following foods: roasted peanuts, corn, and pepperoni pizza? Explain your reasoning for each food.

3. A typical hot dog weighs about 45 grams. Using your data, calculate the number of grams of fat in the hot dog you analyzed? There are about 7 grams of protein in a hot dog. Almost all of the remaining mass is water. Calculate how many Calories are provided by the fat, by the protein, and by the water. What percentage of the Calories in the hot dog can be attributed to the fat?

Name _____   Date _____

Lab Partner _____   Lab Section _____

Data Sheet–Experiment 27

Part I: Percent Fat in Potato Chips or Tortilla Chips

Brand of chips _____

	Sample 1	**Sample 2**
Mass of dish + chips		
Mass of empty dish		
Mass of chips		
Mass of beaker + fat		
Mass of empty beaker		
Mass of fat		
Percent fat in chips		

A simple calculation will allow you to determine if your results agree with the fat content reported by the chip manufacturer. The mass of a potato chip is mostly fat, carbohydrate, protein and water, so if we assume the water content is low, dividing the reported fat content by the sum of masses of the three main components will provide you with an estimated percent fat content of the chips. If the package for your potato chip sample is available, perform this calculation.

Fat content (grams)   _____

Carbohydrate content (grams) _____

Protein content (grams)   _____

Sum (grams)   _____

Percent fat from package = $\dfrac{\text{Fat content (grams)}}{\text{Sum (grams)}}$ × 100% = _____

Do your results agree? If not, why not? Was our assumption about water content a valid one?

Part II: Percent Fat in a Hot Dog (or other meat product)

Description of the sample _____

	Sample 1	Sample 2
Mass of test tube + hot dog		
Mass of empty test tube		
Mass of hot dog		
Mass of container + fat		
Mass of empty container		
Mass of fat		
Percent fat in hot dog		

Conclusion

Your pre-lab question asked you to predict the relative fat contents of the foods analyzed today. How do the actual results compare with your predictions?

How Much Sugar Is in Soft Drinks and Fruit Juices?

INTRODUCTION

According to the World Wildlife Fund (WWF), more than 145 million tons of sugar (sucrose) are produced annually worldwide from sugar cane and beets.[1] This comes at a huge environmental cost due to the heavy use of irrigation water and agricultural chemicals, large amounts of soil degradation and erosion, and runoff that has been blamed for alteration of diverse and fragile ecosystems such as the Everglades in Florida and the Great Barrier Reef in Australia. In this experiment, you will measure the amount of sugar present in a variety of beverages and explore the impact of added sugar on health and the environment. Almost all soft drinks that are not milk-based (i.e. Coke, Pepsi, Sprite, Kool-aid, Gatorade) are essentially sugar solutions with small amounts of additives for flavoring and color. Fruit juices are also mostly sugar solutions with small amounts of other materials, though in this case the sugar is fructose. As you will see, most of these beverages contain a surprisingly large amount of sugar.

Background Information

In this experiment, you will determine the sugar content of various beverages by measuring the density of each, using a simple hydrometer. **Density** is defined as the mass (in grams) per volume (in cubic centimeters or millimeters), of a substance. For instance, the density of water is 1.000 g/mL (or g/cm^3).

A hydrometer is a device designed to quickly measure the density of a liquid. It is based on the fact that floating objects displace their weight in liquid. You may have encountered one when you had your antifreeze checked at your local auto shop. A less-dense floating object displaces less liquid when it floats than a more-dense floating object. Therefore, the less-dense object floats higher in the liquid. When the density of the liquid increases, objects float higher because a smaller volume of liquid needs to be displaced to achieve the same weight of liquid. (One interesting example of this is that you float higher in salt water than you do in fresh water.) A hydrometer is simply a floating object with a scale to measure how high it floats in various liquids.

---

[1] World Wildlife Fund, *Sugar and the Environment: Encouraging Better Management Practices in sugar production.* 2004. Available online at http://assets.panda.org/downloads/sugarandtheenvironment_fidq.pdf

Overview of the Experiment

1. Prepare a hydrometer using a thin-stem plastic pipette.
2. Calibrate your hydrometer using four sugar solutions of known concentration.
3. Use your data to prepare a graph of hydrometer depth versus sugar concentration.
4. Use your hydrometer to investigate several soft-drink samples.
5. Use your graph to determine the sugar concentration in each soft-drink sample.

Pre-lab Question

Water has a density of 1.000 g/mL. Aluminum has a density of 2.7 g/mL. Would you expect aluminum to float or sink in water? Explain why.

EXPERIMENTAL PROCEDURE

I. Constructing a Hydrometer

You will construct a hydrometer of the type shown in *Figure 28.1*.

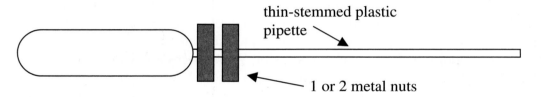

Figure 28.1 A simple homemade hydrometer

1. You will need a <u>thin-stem plastic pipette</u> and one or two <u>metal nuts</u> of the type to fit on machine screws.

2. Put the nut(s) over the stem of the pipette as shown in *Figure 28.1*.

3. Fill the pipette approximately half-full of water.

4. Place the pipette, bulb end down, into a 50-mL graduated cylinder filled nearly to the top with water at room temperature. You need to adjust the amount of water in the pipette so that it floats with the bulb *near* the bottom (but not touching the bottom) and with only a short length of the stem sticking out of the water (as shown in the *Figure 28.2*). If it floats too high or too low, adjust the height by cautiously adding or removing some water from the bulb of the pipette. (Your instructor may give a demonstration of how to do this.) When you have the pipette floating correctly, it is ready to use as a hydrometer.

If the hydrometer is placed into a liquid that is denser than water, it will float with more of its stem out of the water. To determine the relative density, and hence the amount of dissolved solids (mostly sugar) in soft drinks or fruit juices, all you need to do is float the hydrometer in the beverage and measure the length of the stem that sticks out of the liquid. The longer the protruding stem, the denser the liquid, and the greater the sugar content. The reference point for the protruding height will always be the surface of the liquid.

II. Calibrating the Hydrometer

In order to estimate the actual amount of sugar in a soft drink, the hydrometer must be calibrated. First you need to make your reference sugar solutions. Two groups can work together to make 100 mL of each reference solution, with each taking 50 mL for the calibration. You will use sugar cubes to make your reference solutions. Each sugar cube contains 1 tsp of sugar, which has a mass of approximately 4 g. You will be provided with a more accurate mass by your instructor, or you may be asked to weigh the sugar cubes yourself to determine their mass. From this mass, you can calculate the concentration of the reference solutions.

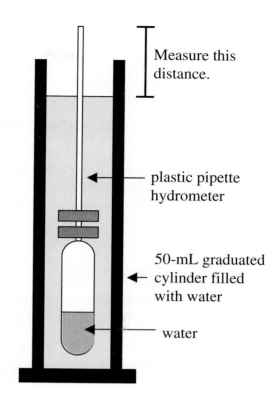

1. Label <u>four 150-mL beakers</u> with the letters A, B, C, and D.

2. Place 80 mL of water in each beaker.

3. Place one sugar cube in beaker A, two sugar cubes in beaker B, three sugar cubes in beaker C, and four sugar cubes in beaker D. Stir each solution until the sugar cubes are dissolved.

Figure 28.2 The hydrometer assembly

4. Transfer the first solution to the <u>100-mL graduated cylinder</u>, and add enough water to make the volume exactly equal to 100 mL. Return the solution to the beaker and stir to thoroughly mix the solution. Repeat this step for the other three solutions. You will use 50 mL of each solution to calibrate your hydrometer.

5. Fill the 50-mL graduated cylinder with pure water. This is the 0% sugar reference solution.

6. Place the hydrometer gently in the cylinder as shown in *Figure 28.2*.

7. Make sure the hydrometer is free floating. To do this, tap it gently once or twice to be sure that it bobs freely and that it does not stick to the sides of the container.

8. Using a <u>ruler</u>, carefully measure the height of the hydrometer stem that sticks out of the fluid, recorded to the nearest millimeter (mm). Tap the hydrometer gently and repeat the measurement. This measurement is a little tricky since it must be from the liquid surface to the exact end of the hydrometer stem. The height variations between liquids will be small, so it is important to measure the heights carefully. (If you are working with a partner, both of you should check the height.) Record the height in the data table.

9. Pour the water out of the graduated cylinder and place approximately 50 mL of your first solution (made with one sugar cube) in the same cylinder. Blot the water off your

hydrometer and use the same procedure to measure the height of the protruding hydrometer stem. Record this height in the data table.

10. Repeat the density measurement for the remaining solutions, going from the least to most concentrated solution. Blot the hydrometer dry between each measurement, and record your measurements in the data table.

III. Testing Soft Drinks and Other Beverages

Make certain that beverages have been decarbonated, or follow any given instructions to do this. If the hydrometer is placed in a carbonated drink, bubbles sticking to the hydrometer will add buoyancy and the reading will be incorrect. The carbonated beverage samples that you will be testing have been decarbonated in advance by boiling for a few minutes and then cooling to room temperature.

Your instructor will specify what samples to test, or you may be offered a variety of samples from which to choose. Members of your class or lab section may be asked to test different samples so that class data can be assembled and compared. The beverage samples will be in graduated cylinders at labeled stations in the lab.

     STOP! Do not drink any beverages that have been opened or used in the laboratory.

1. Using the procedure described in Part II, place your hydrometer in each of the beverage samples and carefully measure the exact height of the protruding stem. *To avoid contaminating the samples, always rinse the hydrometer with pure water and blot it gently before placing it in a new solution.*

2. Record your data in the data table.

Optional: If your instructor permits this, you may take your hydrometer with you (being careful not to squeeze out any of the water) and use it to test other drinks you have in your house or residence hall. It is very important that yoo not drink any samples in which the hydrometer has been immersed. Measurements should be done on small samples and then discarded.

IV. Calculations

The way to make the calibration data more useful is to plot the data as a graph, plotting the hydrometer height (on the vertical axis) against % sugar (on the horizontal axis). You can use a sheet of graph paper or the graph that is provided on the data sheet, which has the horizontal axis already labeled. The vertical axis should be labeled as "height in millimeters." You will need to think carefully about the scales for the vertical axis so that the graph extends over a large part of the paper. *Note that it is NOT necessary for the vertical scale to start at zero.* Enter the data points for the calibration solutions from Part II on the graph (each point has a % sugar value and a height value) by placing a dot in the correct location and drawing a small circle around it so that it is clearly visible. You should have five points on the graph. Connect the points with a smooth line. The line is not necessarily straight; it may be curved, but it should not be drawn as a zig-zag. Therefore you should examine the points carefully before drawing a line.

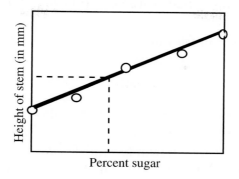

Figure 28.3 Sugar calibration graph

To determine the percent sugar in a soft drink, locate the hydrometer height for that sample on the vertical scale and draw a horizontal line until it intersects the calibration line, as shown above (*Figure 28.3*). From that point, draw a vertical line down the horizontal axis and read the percent sugar from the horizontal scale.

Optional: Use of a computer for graphing. If a computer with a graphing program is available, the data can be plotted, and then the equation for the line can be obtained. With this equation, it is a simple matter to calculate the % sugar from the hydrometer height for each of the beverages.

Sharing Your Results

When the calculations are completed, share your results with other members of the class or lab section. Try to assess the degree of agreement between different student results for the same samples. See what conclusions you can draw about different kinds of beverages.

Clean-up

Discard all solutions as your instructor directs you. Do not put anything down the drain without being explicitly told that it is permitted.

Post-lab Questions

1. Describe how you would modify your hydrometer to increase its accuracy.

2. Explain how the sugar present in fruit juice differs in chemical structure and origin from the sugar you used to make your reference solutions, and from that found in soft drinks and colas.

3. Read the WWF report referenced in the introduction to this experiment.
   a. Identify and discuss one of the major environmental problems with sugar production, and illustrate how it has impacted a specific ecosystem.
   b. Specify a solution to the problem you identified in part (a).

4. Discuss the environmental and health benefits of following a diet that contains foods low in added sugar.

5. In the United States, added sugar in processed foods is mostly in the form of high fructose corn syrup (HFCS), not sucrose. Michael Pollan, author of *In Defense of Food*, recently discussed the environmental issues with production of HFCS with the *Washington Post*.[2] Read the article and discuss the environmental footprint of HFCS.

6. The label on a bottle of regular Pepsi reads:
   Serving size: 8 fluid ounces (about 240 mL)
   Total carbohydrate (as all sugar): 27 g

   a. Calculate to the nearest gram how much sugar you would consume if you drank a 20-ounce bottle of Pepsi.
   b. If a teaspoon of sugar is approximately 4 grams, how many teaspoons of sugar would you have consumed from this bottle of Pepsi?

7. When cans of regular and diet soda are placed in a large container of ice water, some of the cans float while others sink. Which float? Which sink? Explain.

8. Ethanol (an alcohol) is found in beer and wine.
   a. Look up the density of ethanol, and speculate how the density of a mixture of water and ethanol will change with increasing concentration of ethanol.
   b. Devise a method to determine the concentration of ethanol in alcoholic beverages using a hydrometer.
   c. What modification might you have to make to your hydrometer to use it for this measurement?

---

[2] Hartman, E. "High Fructose Corn Syrup: Not So Sweet For The Planet." *Washington Post*, March 9, 2008. Available online at
http://www.washingtonpost.com/wp-dyn/content/article/2008/03/06/AR2008030603294.html

How Much Vitamin C is in Juice and Vitamin Tablets?

INTRODUCTION

Vitamin C is one of the essential vitamins required for good health. Most people obtain the necessary amount of vitamin C in their diet (primarily in fresh fruits, fruit juices, and vegetables), but vitamin C tablets are also widely used as nutritional supplements. In this experiment, you will analyze a sample of fruit juice to find out how much vitamin C it contains and see whether it could provide your daily requirement for vitamin C. You will also analyze a vitamin C tablet.

Background Information

Vitamin C, or ascorbic acid ($C_6H_8O_6$), is chemically similar to the simple sugar, glucose ($C_6H_{12}O_6$), which is plentiful in our bodies. Most animals possess an enzyme needed for making ascorbic acid from glucose, but humans and a few other species lack that enzyme and therefore must secure ascorbic acid directly from foods we consume. The best-known function of ascorbic acid is the prevention of scurvy; a minute amount (10 mg per day) is adequate for this purpose. However, it also plays several other important roles in human health. Most of its functions are related to the fact that it has a strong tendency to transfer electrons to other chemical substances (chemists say that it is a "reducing agent" or "antioxidant"). It acts to prevent other, potentially harmful, reactions that involve the transfer of electrons.

To maintain good health, the current reference daily intake (RDI) for vitamin C is 60 mg. Foods like fruits and vegetables with a high water content often contain large amounts of vitamin C. Vitamin C is very soluble in water, which means that if high doses of vitamin C are ingested, much of it is rapidly excreted in urine.

The **titration method** of analysis has been explained and utilized in several previous experiments and will not be described in detail here. You may wish to refer back to your previous experience with titrations. The titration analysis for ascorbic acid (vitamin C) is based on its tendency to lose electrons. A suitable electron acceptor substance, in this case iodine (I_2), is used as the reagent to "titrate" the ascorbic acid. Although it is not necessary to understand the reaction in order to do the analysis, the chemical equation for the reaction is the following:

$$C_6H_8O_6 + I_2 \rightarrow C_6H_6O_6 + 2\ I^- + 2\ H^+$$

Since it is difficult to prepare iodine solutions with an accurately known concentration, the iodine concentration is determined first by titrating a reference sample of ascorbic acid. The "end point" of the titration is signaled by the appearance of the color of unreacted iodine. The color is made more intense by adding a drop of starch solution, which forms a deep blue color with iodine.

Overview of the Experiment
1. Assemble all of the necessary materials.
2. Titrate the ascorbic acid reference solution.
3. Prepare a solution of a vitamin C tablet in a volumetric flask.
4. Titrate portions of the solution from the vitamin C tablet.
5. Calculate the amount of ascorbic acid in the vitamin tablet.
6. Titrate orange juice or other fruit juice.
7. Calculate the concentration of ascorbic acid in the juice.

Pre-lab Questions
1. Vitamin C is a micronutrient. What does this mean?
2. What is a macronutrient? What are the three main classes of macronutrients that we eat?

EXPERIMENTAL PROCEDURE

I. Preparing the Vitamin C Tablet Sample

1. Weigh a vitamin C tablet and record its mass to the nearest milligram (0.001 gram).

2. Place the tablet in a clean 100-ml volumetric flask. Fill the flask about half-full with water. Shake and swirl the flask until the tablet is broken down. It may be helpful to gently crush the tablet with a glass stirring rod. Label the flask and set it aside to continue dissolving while you go on to the next procedure.

II. Titration of the Ascorbic Acid Reference Solution

1. Using a clean, dry beaker, obtain about 10 mL of the ascorbic acid reference solution. Record its exact concentration (given in mg/mL).

2. Fill a graduated-stem plastic pipet with exactly 1 mL of the reference ascorbic acid solution. (It may take a little practice to fill the pipet exactly to the 1-mL mark.) Then slowly dispense the 1-mL sample into a clean, dry well in the wellplate (e.g., well A1).

3. In the same fashion, add exactly 1 mL of ascorbic acid reference solution to each of 3 or more wells.

4. Add 1 drop of starch solution to each of the four wells.

5. Obtain a small supply of iodine solution in a clean, dry beaker or test tube.

6. Fill the other pipet with iodine solution.

7. Proceed to titrate the first ascorbic acid reference sample with the iodine solution, counting drops and stirring as you go. You will observe a momentary blue color that disappears upon stirring but lingers longer as more iodine is added. The end point occurs when one drop gives a pale blue color that does not disappear upon stirring. Record your data.

Notes on Titrating: You may wish to designate one well in the wellplate for waste. Partial drops can be dispensed into it. Remember that the most consistent drop sizes are obtained by holding the pipet vertically. If you miss the endpoint or lose count of drops, simply go on to another sample. It is easy and fast to titrate additional portions of the solution as needed. If you wish to review a more detailed description of titration, re-read the introduction to Experiment 13.

8. In the same fashion, titrate at least 3 more ascorbic acid reference samples. Record the results on the data sheet.

III. Titration of the Vitamin Tablet Solution

1. By now, the vitamin C tablet in the volumetric flask should be mostly dissolved (except for insoluble "filler" material that is typically starch).

2. Add deionized water to fill the flask up to the bottom of the neck, then *slowly* add more water to bring the water level up the line etched on the neck of the flask. The last milliliter or so should be added one drop at a time. The bottom of the curved meniscus should just touch the line as shown in *Figure 29.1*.

3. Put the cap on the flask securely and mix the contents of the flask thoroughly by repeatedly turning the flask upside down and swirling. (If necessary, hold the cap with a finger.)

4. Take the graduated-stem pipet that was used for the reference solution. Rinse it thoroughly with deionized water (filling and emptying several times) and then rinse it with the solution from the volumetric flask.

5. Use this pipet to add exactly 1 ml of the vitamin tablet solution to each of four clean wells in the well plate (e.g., row B).

6. Add a drop of starch solution to each of these samples.

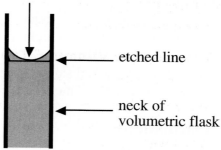

Figure 29.1. Liquid should fill the volumetric flask until the bottom of the meniscus is exactly on the etched line.

7. Using the same iodine solution and plastic pipet that you used in Part II, titrate the first sample to the pale blue endpoint. Record the result.

8. Repeat the titration of the vitamin C tablet solution at least 3 more times and record the results.

IV. Titration of a Juice Sample

1. Rinse the beaker used in Part II thoroughly with deionized water. Also rinse the plastic pipet thoroughly or obtain a new pipet for the juice sample.

2. Wipe the beaker dry; then obtain a sample of orange juice or other juice. If several juice samples are being tested, record which one you used.

 CAUTION! Do not consume any beverage from a container that has been open in the laboratory.

3. Rinse the pipet with the juice sample.

4. Use this pipet to add exactly 1 ml of juice to a fresh well in the wellplate.

5. Add one drop of starch solution.

6. Titrate this juice sample in the same way you titrated the vitamin C tablet solution. The color of the juice affects the appearance of the end point; therefore, it is good idea to do one titration for practice to see what the end point looks like. Depending on the number of drops used, you might choose to use a larger or smaller volume of juice for the subsequent titrations. If so, be sure to record the volumes actually used.

7. Then do at least 3 more titrations of the juice. Remember to add a drop of starch solution to each. Record your data.

V. Optional Extensions or Alternate Assignments

A variety of juices and foods can be analyzed for vitamin C by this basic method. Once you know how to do this analysis and you have done the calibration with a known reference solution, it is easy to analyze additional samples. Your instructor may divide up the class so that different samples will be analyzed by various members of the class. In addition to fruit juices and other beverages, it is possible to extract the liquid from vegetables such as broccoli or peppers (both of which are high in vitamin C) and titrate the extracted liquid.

If you analyze additional samples, record the data on a separate sheet of paper, using the same format as shown on the data sheet.

VI. Calculations

1. Look carefully at the data from the three sets of titrations to see whether the results show consistency or whether any individual result should be eliminated because it appears to be very different from the other data in the set. If so, make a note beside that value and omit it from your calculations.

2 Calculate the average number of drops of iodine used for each set of titrations.

3. For the first set, divide the milligrams of ascorbic acid per milliliter by the average number of drops of iodine used (equation 1). This gives the mg of ascorbic acid corresponding to 1 drop of iodine solution. This is the **calibration factor** for this analysis. It is unique for the particular iodine solution and your equipment.

$$\frac{\text{mg ascorbic acid}}{\text{1 mL of ref. solution}} \times \frac{\text{1 mL of ref. solution}}{\text{drops of iodine}} = \frac{\text{mg ascorbic acid}}{\text{1 drop of iodine}} \qquad \textbf{(1)}$$

4. For the analysis of the vitamin C tablet, multiply the average number of drops of iodine used by the calibration factor to obtain the mg of ascorbic acid per mL (equation 2).

$$\frac{\text{drops of iodine}}{\text{1 mL sample}} \times \frac{\text{mg ascorbic acid}}{\text{1 drop of iodine}} = \frac{\text{mg ascorbic acid}}{\text{1 mL sample}} \qquad \textbf{(2)}$$

   Multiply the answer from equation (2) by 100 to find the total milligrams of ascorbic acid in the 100-mL volumetric flask. This is the same as the milligrams of ascorbic acid in the original tablet.

5. For the analysis of the juice sample, use equation (2) to calculate the milligrams of ascorbic acid per milliliter of juice.

6. If a standard serving of juice is 8 ounces and 1 fluid ounce is 30 mL, calculate how much vitamin C would be obtained from an 8-ounce serving of this juice.

Post-lab Questions

1. Is vitamin C a water-soluble or fat-soluble vitamin? How does this relate to the procedure we used for analysis?

2. What fraction of the mass of the vitamin tablet was ascorbic acid?

3. How closely did your analysis of the vitamin tablet match the label on the bottle? Calculate the percentage difference between them.

4. The reference daily intake (RDI) cited in the introduction is the minimum amount recommended to the general population by the U.S. Food and Drug Administration that will allow most people to maintain health. In 2004, the Institute of Medicine (part of the National Academies) published more detailed Dietary Reference Intake (DRI) values that take into account age, sex, and activity levels of individuals when recommending nutrient intake levels. DRI tables for Vitamin C and other nutrients can be downloaded at the following website:

    http://iom.edu/Activities/Nutrition/SummaryDRIs/DRI-Tables.aspx

    a. Using the tables at the above website, find your DRI for Vitamin C. How much Vitamin C should you take in each day?

    b. How much of the juice that you analyzed would you have to drink to meet the U.S. RDA for vitamin C? (**Note:** some nutritionists think the RDAs for vitamin C are too low and that 200 mg/day is more nearly optimal.)

    c. If you analyzed other foods, how much of each would you have to eat to meet the U.S. RDA for vitamin C?

Notes

Name _____   Date

Lab Partner _____   Lab Section _____

Data Sheet—Experiment 29

Part II: Titrations of the Reference Solution

Concentration of reference ascorbic acid _____ mg/mL

Trial	#1	#2	#3	#4	#5
mL of reference sol'n	1.00	1.00	1.00	1.00	
Drops of iodine					

Note: *Extra columns are for optional additional titrations.*

Average drops of iodine solution = _____ per 1 mL of reference solution

Calibration factor = _____ mg of ascorbic acid per drop of iodine

Part III: Titration of Vitamin Tablet

Mass of tablet: _____

Titration	#1	#2	#3	#4	#5
mL of sample	1.00				
Drops of iodine					

Average drops of iodine solution per 1 mL of sample: _____

mg of ascorbic acid per 1 mL:_____ mg of ascorbic acid per tablet: _____

Part IV: Titration of a Juice Sample

Description of the sample: _____

Titration	#1	#2	#3	#4	#5
mL of sample	1.00				
Drops of iodine					

Average drops of iodine solution per 1 mL of juice: _____

mg of ascorbic acid per 1 mL: _____ mg of ascorbic acid per 8 ounces:_____

Notes

How Can We Isolate DNA?

INTRODUCTION

DNA is in the media frequently in conjunction with topics as diverse as criminal investigations, genetic engineering, and genetically induced diseases. Sections 12.2-12.3 in *Chemistry in Context* discuss DNA in considerable detail. In this experiment, you will isolate and see some DNA, from a banana, an onion, or from your own cheek.

Background Information

Deoxyribonucleic acid (DNA) molecules contain the blueprint for life and are found in the nucleus of any living cell. The makeup of DNA molecules is both simple and complex. It is simple because the molecule that contains the blueprint for a living organism is a polymer consisting of only six chemical fragments (phosphate, deoxyribose, guanine, cytosine, adenine, thymine). It is complex because the phosphate and deoxyribose bond together to form a framework to which the four other fragments are attached. The sequence of these four fragments on the framework is the basis for an organism's genetic code.

A single strand of DNA can have thousands of fragments hooked together in a very specific order. Hydrogen bonding causes the long molecules to assume a helical shape. The helix consists of two strands of DNA wrapped around each other. (See *Figure 12.6(b)* in *Chemistry in Context* for a diagram of the double helix.) Many methods have been developed for extracting DNA from cells. The procedures are designed to extract only the DNA and leave other cell components behind. If you consider the number and types of proteins and lipids found in a cell, it is surprising that DNA is relatively easy to isolate, at least from some kinds of tissue. For example, the DNA in onion cells can be easily isolated by first disrupting the cell walls with a homogenizing solution consisting of detergent, salt, and a reagent to control the pH. The DNA that "leaks" out of the broken cells can be "precipitated" (made insoluble) by addition of ethanol. Because DNA molecules are long polymer chains, they tend to bind together when they are precipitated to form long strands that will mat together and look something like coagulated egg white. The strands can be wrapped around a thin glass or plastic rod and removed. DNA from bananas or the inside of your cheeks is likewise simple to isolate. Your instructor will let you know which type of sample you will use for this experiment.

Overview of the Experiment

1. Prepare your sample of banana, onion, or cheek cells.
2. Homogenize and heat your sample.
3. Mash the onion preparation to break up cells.
4. Chill the mixture, then filter it to remove the cell walls.
5. Precipitate the DNA with alcohol.
6. Spool the DNA around a glass or plastic rod.

Pre-lab Question

DNA is a polymer where each repeat unit consists of a phosphate, sugar and base. Draw a schematic representation of DNA showing how these units are connected to each other. You may wish to consult *Figure 12.4* in *Chemistry in Context*.

EXPERIMENTAL PROCEDURE

A. Banana DNA

1. Obtain a bucket, put a bottle of isopropyl alcohol inside, and fill the rest of the bucket with ice to chill the alcohol.

2. In a blender, combine 10 mL of Woolite detergent, 1/2 teaspoon of NaCl (table salt), and 90 mL of water, at high speed. Once blended, pour the solution into a beaker and place the beaker into a hot water bath (at about 60°C) for approximately 15 min. Then, using a hand mitt carefully remove the beaker from the hot water bath and return it to the blender.

3. Add 1/2 of a banana to the blender and blend it at high speed with the soap/salt solution. Return the banana mixture to the beaker and place the beaker in an ice water bath for 5 minutes to cool it down.

4. Carefully place 2 pieces of cheesecloth over a funnel placed in a beaker. This will filter the mixture and remove any unmashed pieces of banana.

5. Place 6 mL of banana filtrate into a test tube. Then, tilt the tube and carefully pour 9 mL of cold isopropyl alcohol down the side. Try to prevent the alcohol from mixing with the banana mixture.

6. Allow the isopropyl alcohol to sit atop the banana mixture for five minutes without disturbing the test tube. Bubbles will begin to form and rise up from the bottom of the test tube to the surface of the isopropyl alcohol. The banana DNA will begin to rise and precipitate out of solution. The DNA will be cloudy white and is extremely fragile, so take care to avoid any sudden disturbance to the test tube.

7. Gently place a glass stirring rod or a toothpick into the test tube without disturbing the solution. Carefully swivel it and watch the DNA wrap around the surface.

B. Onion DNA

1. Weigh 10 grams of chopped onion into a large test tube.

2. Add 20 mL of "homogenizing solution" (see Background section) and place the test tube in a 60°C water bath for 10 minutes.

3. Keep the onion mixture in the beaker of 60°C water for 10 minutes. It is important to monitor the temperature and be sure that it stays close to 60°C. If the temperature of the onion mixture reaches 70°C, the DNA structure will be destroyed, and you will not be able to isolate and spool it. (**Note:** Incubating with the homogenizing solution serves to soften the cell walls, break down the protein in the cells, and may also help to dissolve cell membranes.)

4. After the onion mixture has heated for 10 minutes, put the test tube in an ice-water bath and chill until the mixture reaches 15°C.

5. Pour the cooled solution into a mortar and use the pestle to mash the onion to a smooth paste.

6. Transfer the paste and all the liquid to a 50-mL beaker and chill in an ice bath for 15 minutes.

7. Using a funnel, filter the mixture through coarse filter paper into a 50-mL graduated cylinder and return the liquid to the ice bath for 5 minutes.

8. Slowly pour 20 mL of 95% ethanol down the side of the graduated cylinder to form a layer on top of the DNA-containing liquid. A white stringy precipitate should form at the liquid/liquid interface.

9. Slowly pour the liquid into a small Petri dish. Then put the end of a thin glass rod or a glass Pasteur pipet into the liquid and slowly twirl it around. White strands of DNA should wrap themselves around the end of the pipet.

C. DNA from Human Cheek Cells

1. Pour 10 mL of fresh tap water or bottled water into a clean, 30 mL plastic drinking cup. Put the water into your mouth and swirl for at least 30 seconds. Swirling the water for longer is better. Spit the water back into the plastic cup. This swirling washes cells from the inside of your cheeks into the water.

2. Add 1 mL of 8% NaCl solution to a large test tube, and to this add 5 mL of the "cheek cell" water.

3. To the same tube, add 1 mL of 10% sodium lauryl sulfate solution **or** 1 mL of 25% liquid dishwashing detergent solution. This detergent ruptures the cell membranes to release the DNA into the salt solution.

4. Place the test tube into a 55°C water bath for 5 minutes. This enhances the action of the detergent and also denatures enzymes that might damage the DNA.

5. Remove the tube from the water bath, place a stopper on top, and mix the contents by gently inverting the tube several times. Do not shake the tube.

6. To a new test tube, add 1 mL of 95% ethyl or isopropyl alcohol. Place this tube in a beaker full of ice to chill it.

7. Place a clean glass stirring rod into the test tube that contains the "cheek cell" mixture. Transfer a small amount of the solution into the test tube containing the alcohol. Observe the DNA strands floating in the alcohol.

Clean-up

Dispose of liquid and solid waste in appropriate containers as specified by your instructor. Clean and dry all glassware.

Post-lab Questions

1. In this experiment, you were cautioned not to allow the temperature to go above 60°C. Why do you think DNA decomposes when it is heated to too high a temperature? (**Hint:** Think about the double-helix structure.)

2. DNA was precipitated out of a water solution by adding ethanol (CH_3-CH_2-OH) or isopropanol. Why do you think DNA is less soluble in ethanol than in water? Remember that ethanol is partly a hydrocarbon and therefore is a less polar molecule than water.

3. Below are the structural formulas for the four nucleotide bases in DNA: adenine, thymine, cytosine, and guanine. What features do they have in common?

Adenine   Thymine   Cytosine   Guanine

4. Which hydrogen atoms in the 4 nucleotides are responsible for forming the hydrogen bonds that hold DNA strands together in the double helix? How <u>many</u> hydrogen bonds are possible for each? See *Figure 12.7* of *Chemistry in Context* for help. Suggest an explanation for why adenine always pairs with thymine and cytosine always pairs with guanine.

5. In light of your answer to Question 4, suggest why DNA molecules that are rich in cytosine and guanine will "melt" or decompose at a higher temperature than DNA that is rich in adenine and thymine.

6. The diversity of life depends on the fact that a great many combinations of the 4 nucleotide bases (shown above) can be assembled into long chains. Illustrate this by writing out 20 of the 64 possible combinations of just 3 bases, using any combinations of the four building blocks (designated as A, T, C, and G).

Name _____   Date _____

Lab Partner _____   Lab Section _____

Data Sheet–Experiment 30

Sample used: _____

Observations

Describe the appearance of the mixture prepared with your sample and the detergent/homogenizing solution, and any changes you observe as the reaction occurs.

Describe what happens as the solution comes into contact with the ethanol or isopropanol.

Describe the appearance of the DNA you isolate.

Notes

Performance-Based Assessment Activities

General Instructions: Your instructor may assign one or more of these activities as a test of whether you have learned how to carry out scientific investigations. First, write up a brief description of your experimental plan and check it with your instructor for feasibility and safety, then perform the experiment. You will need to keep a careful written record of your method and your observations. Upon completion of the experiment, you will be asked to describe your procedure and explain how your conclusions are based on your recorded observations.

1. A spectrophotometric study to follow Experiment 5

   Design and test a method for determining the relationship between the concentration of a colored solution and its absorption of light. Include in your procedure a way to determine the concentration of an unknown solution.

2. A quantitative chemical mole study to follow Experiment 8

   When calcium carbonate reacts with hydrochloric acid, there are three possible products: $CaCl_2$, $CaCl_2 \cdot 2H_2O$, and $CaCl_2 \cdot 6H_2O$. Design and carry out an experiment to find out which product is formed.

3. A calorimetry study to follow Experiment 10

   Design and carry out an experiment to measure the energy content of wood. (A suggested source of wood is the wood splints that are commonly available in chemistry laboratories.) First identify the experimental challenges and devise a way to solve them.

4. A calorimetry study to follow Experiment 10

   Using nuts as an example of a high-calorie food, design and carry out an experiment to measure the energy content in nuts and compare it with other fuels. First identify the experimental challenges and devise a way to solve them.

5. A comparison study of volume vs. weight for titrations, to follow Experiment 13, 14, 15, or 16

   Using a titration analysis of your choice, design and carry out an experiment to compare titrations done by volume (i.e., by counting drops) with titrations done by weight (i.e., by weighing the pipets on a laboratory balance). What are the advantages and disadvantages of each method?

6. <u>A study of melting points to follow Experiment 26</u>

Design and carry out an experiment to test whether melting point is an accurate measure of purity. Starting with pure samples of aspirin and salicylic acid, determine how melting point varies in mixtures and whether melting point can be used to determine composition of such a mixture.

7. <u>A density of solutions study to follow Experiment 28</u>

Develop and test a procedure for measuring the alcohol content in an unknown alcohol-water mixture. Pure alcohol has a density that is lower than that of water.

8. <u>A density of solutions study to follow Experiment 28</u>

Devise and test a procedure for measuring the ethylene glycol content in automobile radiator coolant. Pure ethylene glycol has a higher density than water. (Most automobile coolant systems contain ethylene glycol mixed with water. This prevents freezing in winter as well as allowing the engine to run at a higher temperature than with pure water.)

9. <u>A study of vitamin C stability to follow Experiment 29</u>

Design and carry out an experiment to test whether vitamin C in orange juice decomposes during storage and whether temperature makes a difference.